Lecture Notes in Mathematics

Edited by A. Dold and B. Eckmann

793

Jean Renault

A Groupoid Approach to C*-Algebras

Springer-Verlag
Berlin Heidelberg New York 1980

Author

Jean Renault
Département de Mathématiques
Faculté des Sciences
45 Orléans – La Source
France

AMS Subject Classifications (1980): 22 D 25, 46 L 05, 54 H 15, 54 H 20

ISBN 3-540-09977-8 Springer-Verlag Berlin Heidelberg New York
ISBN 0-387-09977-8 Springer-Verlag New York Heidelberg Berlin

Printing and binding: Beltz Offsetdruck, Hemsbach/Bergstr.
2141/3140-543210

CONTENTS

INTRODUCTION

The interplay between ergodic theory and von Neumann algebra theory goes back to the examples of non-type I factors which Murray and von Neumann obtained by the group measure construction [54]. A natural and probably definitive point of view which joins both theories has recently been exposed by P. Hahn [45]. It uses the notion of measure groupoid, introduced by G. Mackey "to bring to light and exploit certain apparently far reaching analogies between group theory and ergodic theory" ([53], p.187). In particular, the group measure algebra may be regarded as the von Neumann algebra of the regular representation of some principal measure groupoid. Moreover, most of the properties of the algebra may be interpreted in terms of the groupoid. The same standpoint is adopted by J. Feldman and C.Moore [31], in the framework of ergodic equivalence relations. Besides, they characterize abstractly the von Neumann algebras arising from their construction.

It is natural to expect that topological locally compact groupoids play a similar role in the theory of C^*-algebras. The notions of topological and of Lie groupoid were introduced by Ehresmann for applications to differential topology and geometry. More recent interest in topological groupoids has come from the theory of foliations ([10] ,p.273). It seems to be the differential geometry point of view, rather than Mackey's virtual group point of view which aroused J. Westman's interest in groupoids and led him to the construction of convolution algebras of groupoids, first in the transitive (and locally trivial) case [75] and then in the non-transitive principal case [77]. However the relevance to the theory of induced representations is also apparent in [75]. Convolution algebras of transformation groups had already been used for some time [16,37]. The main works about transformation group C^*-algebras, by Effros and Hahn [24] and by Zeller-Meier[80], appeared at about the same time as Westman's article. Although their main purpose is to construct interesting examples of C*-algebras, Effros and Hahn also give some results on the structure of a transformation group C*-algebra. This goal is more apparent in Zeller-Meier's work, which is more directly motivated by group representation theory. Most of the latter work about transformation group C*-algebras concerns itself with the structure theory of these algebras (for example [39]).

The starting point of this work is a theorem of S. Strǎtilǎ and D.Voiculescu about approximately-finite (or AF) C*-algebras [69]. Generalizing the method of L.Garding and A. Wightman [34] for studying factor representations of the canonical anticommutation relations of mathematical physics, they show that every AF C*-algebra can be diagonalized and use a diagonalization to study its structure and its representations. In our setting, this amounts to saying that every AF C*-algebra is the C*-algebra of a principal groupoid (3.1.15).

The construction (2.1) of the C*-algebra of a groupoid is modelled after the construction of the C*-algebra of a transformation group given by Effros and Hahn. Since a locally compact groupoid does not necessarily have a Haar system, (Westman uses the term of left invariant continuous system of measures), needed to define the convolution product, and since such a Haar system need not be unique, (although some results about existence and uniqueness of Haar systems can be found in K. Seda's articles [67,68], we consider locally compact groupoids with a fixed Haar system. The case of r-discrete groupoids, which generalize discrete transformation groups, deserves special attention, because it includes all our examples. An r-discrete groupoid has a Haar system if and only if its range map is a local homeomorphism, and, if this is the case, it is a scalar multiple of the counting measures system (1.2.8.). In the general case, but under suitable hypotheses, we show that the strong Morita equivalence class of the C*-algebra does not depend on the choice of the Haar system (2.2.11).

The theory of group C*-algebras suggests many generalizations. In particular, one expects a correspondence between unitary representations of the groupoid and non-degenerate representations of its C^*-algebra. This is established (2.1.23) under a rather technical condition which will often be needed, namely the existence of sufficiently many non-singular Borel G-sets (definition 1.3.27). It is also possible to induce a representation from a closed subgroupoid (2.2.9). We give a definition of amenability in section 3 of chapter 2. It develops that the C^*-algebra of an amenable groupoid concides with the reduced C^*-algebra, obtained by considering only the representations induced from the unit space (2.3.2). Moreover, using some of R. Zimmer's ideas about amenable measure groupoids [82,83], it is easily shown that this C^*-algebra is nuclear (2.3.5).

From our point of view, the most interesting groupoids are principal groupoids. Their C^*-algebras appear as genuine generalizations of matrix algebras. We have looked for a characterization of these algebras similar to the condition given by Feldman and Moore for algebras over an ergodic equivalence relation. The notion of Cartan subalgebra we give (2.4.13) is rather restrictive and not as congenial as the corresponding notion for von Neumann algebras. In particular, we show by an example (3.1.17) that a regular maximal self-adjoint abelian subalgebra which is the image of a unique conditional expectation need not be a Cartan subalgebra. The correspondence between closed two-sided ideals of the reduced C^*-algebra of a principal groupoid and the closed invariant subsets of its unit space is established in the r-discrete case (2.4.6).

A continuous homomorphism (also called a one-cocycle) from a locally compact groupoid to a locally compact abelian group defines a continuous homomorphism of the dual group into the automorphism group of the C^*-algebra of the groupoid (2.5.1). Moreover many one-parameter automorphism groups of the AF C^*-algebras considered in mathematical physics (e.g. gauge automorphism group, dynamical groups) arise in this fashion (examples 3.1.6 and 3.1.10). The groupoid point of view is particularly well suited to their study. For example, the Connes spectrum of such an automorphism group is the asymptotic range of the cocycle (2.5.8) and the crossed product C^*-algebra is the C^*-algebra of the skew-product (2.5.7). Besides, the KMS condition for states may be replaced by a condition much closer to the original Gibbs Ansatz characterizing equilibrium states (2.5.4). We use groupoids to derive particular but important cases of some theorems of D. Olesen and G.K. Pedersen [58] about simplicity and primitivity of crossed product C^*-algebras as well as the main results of O.Bratteli [9]. Another application of groupoid C^*-algebras is the study of the C^*-algebra of the bicyclic semi-group and of the Cuntz C^*-algebras (3.2).

A number of fundamental problems have not been touched in this work. As we have seen earlier, groupoids have been introduced for two reasons. One is the "virtual group" point of view, we have not even given a definition of similarity for locally compact groupoids with Haar system. The other is the application to differential geometry, in particular to the theory of foliations ; we have not made any mention of the work of A. Connes in this direction. These topics must await further development in the future.

The author wishes to express indebtness to Marc Rieffel for numerous and fruitful suggestions and to Paul Muhly for a careful reading of the manuscript. He would like to thank P. Hahn, who taught him about groupoid algebras and J.Westman for some unpublished material he gave him.

CHAPTER 1

LOCALLY COMPACT GROUPOIDS

The first chapter sets up the framework of this study. To gain some motivation for the definitions which are given there, the reader can look simultaneously at the examples of the third chapter. It is also useful to keep in mind the example of transformation groups, which is recalled below, and which suggests most of the terminology.

The first section gives the algebraic setting of the theory. The two main concepts are groupoids and inverse semi-groups. The definition of a locally compact groupoid with Haar system is introduced in the second section. The third section deals with the notion of quasi-invariant measure, and a generalization of it, the KMS condition. The results thereof will be of great use in the second chapter. Some elementary properties of one-cocycles are studied in the fourth section. Given a one-cocycle, one can build the skew-product groupoid, and a basic question is to determine its structure in terms of the cocycle. An essential tool is the asymptotic range of the cocycle, which is the topological analog of Krieger's asymptotic ratio set in ergodic theory (see [31], I, definition 8.2).

1. Definitions and Notation

We shall use the definition of a groupoid given by P. Hahn in [44] (definition 1.1). It is essentially the same as the one used by J.Westman in [79] and the one used by A. Ramsay in [61].

1.1. Definition : A groupoid is a set G endowed with a product map $(x,y) \mapsto xy$: $G^2 \to G$ where G^2 is a subset of G x G called the set of composable pairs, and an inverse map $x \to x^{-1} : G \to G$ such that the following relations are satisfied :

(i) $(x^{-1})^{-1} = x$

(ii) $(x,y), (y,z) \in G^2 \Rightarrow (xy,z),(x,yz) \in G^2$ and $(xy)z = x(yz)$

(iii) $(x^{-1},x) \in G^2$ and if $(x,y) \in G^2$, then $x^{-1}(xy) = y$

(iv) $(x,x^{-1}) \in G^2$ and if $(z,x) \in G^2$, then $(zx)x^{-1} = z$

If $x \in G$, $d(x) = x^{-1}x$ is the domain of x and $r(x) = xx^{-1}$ is its range. The pair (x,y) is composable iff the range of y is the domain of x. $G^0 = d(G) = r(G)$ is the unit space of G, its elements are units in the sense that $xd(x) = x$ and $r(x)x = x$. Units will usually be denoted by u, v, w while arbitrary elements will be denoted by x, y, z.

If A and B are subsets of G, one may form the following subsets of G :

$A^{-1} = \{x \in G : x^{-1} \in G\}$

$AB = \{z \in G : \quad x \in A, y \in B : z = xy\}$.

A groupoid G is said to be principal if the map (r,d) from G into $G^0 \times G^0$ is one-to-one, it is said to be transitive if the map (r,d) is onto.

For $u, v, \in G^0$, $G^u = r^{-1}(u)$, $G_v = d^{-1}(v)$, $G^u_v = G^u \cap G_v$ and $G(u) = G^u_u$, which is a group, is called the isotropy group at u.

The relation $u \sim v$ iff $G^u_v \neq \emptyset$ is an equivalence relation on the unit space G^0. Its equivalence classes are called orbits and the orbit of u is denoted $[u]$. G^0/G denotes the orbit space. A groupoid is transitive iff it has a single orbit.

1.2. Examples :

a. Transformation groups

Suppose that the group S acts on the space U on the right. The image of the point u by the transformation s is denoted $u\cdot s$. We let G be $U \times S$ and define the following groupoid structure : (u,s) and (v,t) are composable iff $v = u\cdot s$, $(u,s)(u\cdot s,t) = (u,st)$, and $(u,s)^{-1} = (u\cdot s,s^{-1})$. Then $r(u,s) = (u,e)$ and $d(u,s) = (u\cdot s,e)$. The map $(u,e) \mapsto u$ identifies G^0 with U. The terminology of orbits comes from this example. Then G is principal iff S acts freely, and transitive iff S acts transitively.

b. The groupoid G^2

The set G^2 of composable elements may be given the following groupoid structure : (x,y) and (y',z) are composable iff $y' = xy$, $(x,y)\,(xy,z) = (x,yz)$, and $(x,y)^{-1} = (xy,y^{-1})$.

Then $r^2\,(x,y) = (x,r(y)) = (x,d(x))$ and $d^2\,(x,y) = (xy,d(xy))$. The map $x \mapsto (x,d(x))$ identifies the unit space of G^2 with G. The groupoid G^2 is principal. One may notice that it comes from the action of G on itself. It is transitive iff G is a group.

c. Equivalence relations

Let R be the graph of an equivalence relation on a set U. We give to R the following groupoid structure : (u,v) and (v',w) are composable iff $v' = v$, $(u,v)\,(v,w) = (u,w)$, and $(u,v)^{-1} = (v,u)$. Then, $r(u,v) = (u,u)$ and $d(u,v) = (v,v)$. The unit space of R is the diagonal and may be identified with U. R is a principal groupoid. Conversely, if G is a principal groupoid, (r,d) identifies G with the graph of the equivalence relation $\sim$.

d. Group bundle

A group bundle G is a groupoid such that fo any $x \in G$, $d(x) = r(x)$. A group bundle is the union of its isotropy groups $G(u)$. Here, two elements may be composed iff they lie in the same fiber. Given any groupoid G, $G' = \{x \in G \;:\; d(x) = r(x)\}$ is a group bundle. We call it the isotropy group bundle of G. It is reduced to the unit space of G iff G is principal.

1.3. Definition : Let G and H be groupoids. A map $\phi : G \to H$, is a homomorphism if for any $(x,y) \in G^2$, $(\phi(x), \phi(y)) \in H^2$ and $\phi(x)\,\phi(y) = \phi(xy)$. Then $\phi(u) \in H^0$ if $u \in G^0$. $\phi^0 : G^0 \to H^0$ denotes the restriction of ϕ to the unit spaces. $\phi^2 : G^2 \to H^2$ is the map $\phi^2(x,y) = (\phi(x),\phi(y))$; it is a homomorphism. Two homomorphisms $\phi,\psi : G \to H$ are similar (write $\phi \sim \psi$)if there exists a function $\theta : G^0 \to H$ such that $(\theta\circ r)(x)\,\phi(x) = \psi(x)\,(\theta\circ d)(x)$ for any $x \in G$. Groupoids G and H are called similar (write $G \sim H$) if there exist homomorphisms $\phi : G \to H$ and $\psi : H \to G$ such that $\phi \circ \psi$ and $\psi \circ \phi$ are similar to identity isomorphisms.

Before giving a result of Ramsay [61] (theorem 1.7, p. 260) which illustrates this notion, we need a definition.

1.4. Definition : Let G be a groupoid, E a subset of G^0 ; $G_{|E} = \{x \in G : r(x) \in E$ and $d(x) \in E\}$ is a subgroupoid of G with unit space E ; $G_{|E}$ is called the reduction of G by E.

1.5. Proposition : Let G be a groupoid, E a subset of G^0 which meets each orbit in G^0 ; then $G_{|E} \sim G$.

1.6. Definition : Let G be a groupoid, A a group and $c : G \to A$ a homomorphism, the skew-product G(c) is the groupoid G x A where : (x,a) and (y,b) are composable iff x and y are composable and $b = ac(x)$, $(x,a)(y,ac(x)) = (xy,a)$, and $(x,a)^{-1} = (x^{-1},ac(x))$; $r(x,a) = (r(x),a)$, $d(x,a) = (d(x),ac(x))$. Its unit space is $G^0 \times A$.

A basic example of skew-product is the following. Let s be a transformation of the space U into itself and let f be a function on U with values in an abelian group A. On the space U x A, define the transformation t by $(u,a)t = (us,a + f(u))$. Let us define the groupoid G of s as the groupoid associated with the corresponding transformation group $(U, \mathbb{Z})$ and define similarly the groupoid of t. We leave to the reader to check that the groupoid of t is the skew-product of the groupoid G of s by the homomorphism $c : G \to A$ obtained from f by the rules

$$c(u,n) = \sum_{0}^{n-1} f(ut^i) \text{ for } n \geq 1,$$

$$c(u,0) = 0, \text{ and}$$

$$c(u,-n) = -c(u,n) \text{ for } -n \leq -1.$$

Another important way of building up new groupoids from old ones is the semi-direct product.

1.7. Definition : Let G be a groupoid, let A be a group and let $\alpha : A \to \mathrm{Aut}(G)$ be a homomorphism. We write $x \cdot a = [\alpha(a^{-1})](x)$ for $a \in A$ and $x \in G$. The semi-direct product $G \times_\alpha A$ is the groupoid G x A where (x,a) and (z,b) are composable iff $z = y \cdot a$

with x and y composable, $(x,a)(y \cdot a,b) = (xy,ab)$, and $(x,a)^{-1} = (x^{-1} \cdot a, a^{-1})$. Then, $r(x,a) = (r(x),e)$ and $d(x,a) = (d(x) \cdot a,e)$. The unit space may be identified with G^0.

An example of semi-direct product is the groupoid associated with a transformation group (U,A). In this case $G = U$ is reduced to its unit space. When G is a group, 1.7 is the usual notion of semi-direct product.

There is a natural action of A on the skew-product $G(c)$, namely the homomorphism α defined by the formula $\alpha(a)(x,b) = (x,ab)$ and there is a natural homomorphism c of the semi-direct product $G \times_\alpha A$ into A, defined by the formula $c(x,a) = a$.

1.8. Proposition : With above notation,

(i) $G(c) \times_\alpha A$ is similar to G and

(ii) $(G \times_\alpha A)(c)$ is similar to G.

Proof : One may apply 1.5. For example, to prove (i), one observes that the subset $E = G^0 \times \{e\}$ of the unit space $G^0 \times A$ of $G(c) \times_\alpha A$ meets each orbit. For further reference, let us write down explicitly the similarity homomorphisms :

(i) Define ϕ from $G(c) \times_\alpha A$ to G by $\phi(x,a,b) = x$, define ψ from G to $G(c) \times_\alpha A$ by $\psi(x) = (x,e,c(x))$ and define θ from $G^0 \times A$ to $G(c) \times_\alpha A$ by $\theta(u,a) = (u,e,a^{-1})$ and check that $\phi \circ \psi\ (x) = x$ and $\theta[r(x,a,b)](x,a,b) = \phi\circ\psi\ (x,a,b)\ \theta[d(x,a,b)]$.

(ii) Define ϕ from $(G \times_\alpha A)(c)$ to G by $\phi(x,a,b) = x \cdot b^{-1}$, define ψ from G to $(G \times_\alpha A)(c)$ by $\psi(x) = (x,e,e)$ and define θ from $G^0 \times A$ to $(G \times_\alpha A)(c)$ by $\theta(u,a) = (u \cdot a^{-1},a,e)$ and check that $\phi \circ \psi(x) = x$ and $\theta[r(x,a,b)](x,a,b) = \psi \circ \phi(x,a,b)\ \theta[d(x,a,b)]$.

Q.E.D.

Together with the notion of groupoid, the notion of inverse semi-group plays an important role in this work. The definition given below, as well as some elementary properties, can be found in [11], page 28, or [1].

<u>1.9. Definition :</u> An <u>inverse semi-group</u> is a set $\mathcal{G}$ endowed with an associative binary operation, noted multiplicatively, and an inverse map $s \to s^{-1} : \mathcal{G} \to \mathcal{G}$ such that the following relations are satisfied: $ss^{-1}s = s$ and $s^{-1}ss^{-1} = s^{-1}$.

Then the inverse map is an involution. If $s \in \mathcal{G}$, $d(s) = s^{-1}s$ is the domain of s and $r(s) = ss^{-1}$ is its range. The set of idempotent elements is denoted by $\mathcal{G}^0$. Two idempotent elements commute. The relation $e \leq f$ iff $ef = e$ is an order relation on $\mathcal{G}^0$ which makes it into an inf semi-lattice.

The relation between groupoids and inverse semi-groups is given by introducing the notion of G-set of a groupoid.

<u>1.10 . Definition :</u> Let G be a groupoid. A subset s of G will be called a <u>G-set</u> if the restrictions of r and d to it are one-to-one. Equivalently, s is a G-set iff ss^{-1} and $s^{-1}s$ are contained in G^0.

Let $\mathcal{G}$ be the set of G-sets of G. We note that $s,t \in \mathcal{G} \Rightarrow st \in \mathcal{G}$ and $s \in \mathcal{G} \Rightarrow s^{-1} \in \mathcal{G}$. These operations make $\mathcal{G}$ into an inverse semi-group. Note that the notations $d(s)$ and $r(s)$ agree with the previous ones.

A G-set s defines various maps as follows :

(i) for x on G with $d(x) \in r(s)$, the element xs of G is defined by $\{xs\} = \{x\}s$ (this makes sense) ;

(ii) for x in G with $r(x) \in d(s)$, the element sx of G is defined by $\{sx\} = s\{x\}$;

(iii) for u in $r(s)$, the element $u \cdot s$ in $d(s)$ is defined by $u \cdot s = d(us)$. These notations will be used systematically. The map $u \mapsto u \cdot s : r(s) \to d(s)$ will be called the G-map associated with the G-set s. The reader should not have any trouble to check that

$$x(st) = (xs)t \; ; \qquad (ts)x = t(sx) \; ; \qquad (xs)^{-1} = s^{-1}x^{-1} \; ;$$

$$u \cdot (st) = (u \cdot s) \cdot t$$

where, with our convention, $x(st)$ is defined by $\{x(st)\} = \{x\}st$ for the G-sets s and t and similarly $(ts)x$ is defined by $\{(ts)x\} = ts\{x\}$.

To help understanding what G-sets mean, let us look at the case of a transformation group (U,S). Any element s of the group S defines the following G-set of the associated groupoid G : $s = \{(u,s) : u \in U\}$. Its domain and its range are U. The associated G-map is the transformation $u \mapsto u \cdot s$ and there is no ambiguity in the notations. The map from S to the set of G-sets above defined is an inverse semi-group homomorphism. It is one-to-one but usually not onto. Note that in the case of a group, that is, when U is reduced to one point, the G-sets are exactly the elements of S.

J. Westman has developed in [79] a cohomology theory for groupoids which extends the usual group cohomology theory ; it is reproduced here.

Suppose that $\mathcal{C}$ is some category. A map p from a set A onto a set A^0 such that each fiber $p^{-1}(u)$ is an object of $\mathcal{C}$ will be called a $\mathcal{C}$-bundle map and A will be called $\mathcal{C}$-bundle. For example, a group bundle in the sense of 1.2.d is a $\mathcal{C}$-bundle where $\mathcal{C}$ is the category of groups and any such $\mathcal{C}$-bundle is a group bundle. Let A be a $\mathcal{C}$-bundle with bundle map $p : A \to A^0$. Write $A_u = p^{-1}(u)$. Iso(A) = {isomorphisms $\phi_{u,v} : A_v \to A_u$: $u,v, \in A^0$} has a natural structure of groupoid : $\phi_{u,v}$ and $\psi_{v',w}$ are composable iff $v' = v$; then their product is $\phi_{u,v} \circ \psi_{v,w}$, and $\phi^{-1}_{u,v}$ is the isomorphism inverse of $\phi_{u,v}$. The bijection $id_{u,u} \mapsto u$ identifies the unit space of Iso(A) and A^0. Iso(A) is called the isomorphism groupoid of the $\mathcal{C}$-bundle A.

1.11. Definition : Let G be a groupoid. A G-bundle (A,L) is a $\mathcal{C}$-bundle A together with a homomorphism $L : G \to \text{Iso}(A)$ such that $L^0 : G^0 \to A^0$ is a bijection. (We will often identify G^0 and A^0). When $\mathcal{C}$ is the category of abelian groups, one speaks of a G-module bundle.

Given a G-module bundle (A,L), one can form the following cochain complex. Let us first define G^n for any $n \in \mathbb{N}$. The sets G^0, $G^1 = G$ and G^2 have already been defined. For $n \geq 2$, G^n is the set of n-uples $(x_0,\ldots,x_{n-1}) \in G\text{x}\ldots\text{x}G$ such that for $i = 1,\ldots,n-1$, x_i is composable with its left neighbor. A n-cochain is a function f from G^n to A which satisfies the conditions

(i) $p \circ f(x_0,\ldots,x_{n-1}) = r(x_0)$ and

(ii) if $n > 0$ and for some $i = 0,\ldots,n-1$, $x_i \in G^0$, then $f(x_0,\ldots,x_i,\ldots,x_{n-1})$

εA^0. The set $C^n(G,A)$ of n-cochains is an abelian group under pointwise addition. The sequence $0 \to C^0(G,A) \to C^1(G,A) \to \ldots \to C^n(G,A) \xrightarrow{\delta^n} C^{n+1}(G,A) \to \ldots$, where $\delta^0 f(x) = L(x)\, f\circ d(x) - f\circ r(x)$ and $\delta^n(f(x_0,\ldots,x_n)) = L(x_0)f(x_1,\ldots,x_n) + \sum_{i=1}^{n} (-1)^i f(x_0,\ldots,x_{i-1}x_i,\ldots,x_{n-1}) + (-1)^{n+1} f(x_0,\ldots,x_{n-1})$ for $n > 0$, is a cochain complex.

1.12. Definition : The group of n-cocycles of this complex will be denoted by $Z^n(G,A)$, the group of n-coboundaries will be denoted by $B^n(G,A)$ and the n-th cohomology group $Z^n(G,A)/B^n(G,A)$ will be denoted by $H^n(G,A)$.

A section for a G-bundle (A,L) is a function f from A^0 to A such that $p\circ f(u) = u$, where p is the bundle map. A section f is said to be invariant if $L(x)\, f\circ d(x) = f\circ r(x)$ for every $x \in G$. The set of sections will be denoted by $\Gamma(A)$ and the set of invariant sections by $\Gamma_G(A)$. If (A,L) is a G-module bundle, $C^0(G,A) = \Gamma(A)$ and $H^0(G,A) = \Gamma_G(A)$.

A one-cocycle $c \in Z^1(G,A)$ is a one-cochain f from G to A which satisfies $f(xy) = L(x)f(y) + f(x)$. In particular, if A is a constant bundle, that is, each fiber A_u is equal to a fixed abelian group B, and if G acts trivially on A, that is, $L(x)$ is the identity map of B for every x, a one-cocycle $f \in Z^1(G,A)$ is a homomorphism of G into B. In the case of a constant bundle A as above with trivial action, we write $Z^1(G,B)$ instead of $Z^1(G,A)$.

We may also consider one-cocycles with values in a not necessarily abelian group. In this case, (A,L) is a G-bundle where A is a group bundle. We define $Z^1(G,A) = \{f : G \to A : f(xy) = f(x)[L(x)f(y)]\}$, $B^1(G,A) = \{f : G \to A$: there exists $b : G^0 \to B$ such that $f(x) = [b\circ r(x)]^{-1}[L(x)b\circ d(x)]\}$ and the equivalence relation on $Z^1(G,A)$: $f \sim g$ iff there exists $b : G^0 \to B$ such that $f(x) = [b\circ r(x)]^{-1}\ (x)\ [L(x)b\circ d(x)]$.

As for groups, two-cocycles are related to groupoid extensions :

1.13. Definition : Let (A,L) be a G-module bundle, noted multiplicatively. An extension of A by G is an exact sequence of groupoids

$$A^0 \to A \xrightarrow{i} E \xrightarrow{j} G \to G^0 \quad \text{(we also write } (E,i,j))$$

compatible with the action of G on A, in the sense that there exists a section k for j such that

(i) $k(u) = u$ (A^0, E^0 and G^0 are identified)

(ii) $k(x)\, i(a)\, k(x)^{-1} = i(L(x)a)$ for any $(a,x) \in A \times G$ with $p(a) = d(x)$.

Two extensions (E,i,j) and (E',i',j') are equivalent if there exists an isomorphism $\phi : E \to E'$ such that $i' = \phi \circ i$ and $j = j' \circ \phi$. The set of equivalent classes of extensions with the Baer sum is an abelian group denoted $\mathrm{Ext}(A,G)$.

<u>1.14. Proposition :</u> $H^2(G,A) = \mathrm{Ext}(A,G)$.

<u>Sketch of the proof</u> : Given $\sigma \in Z^2(G,A)$, let $E_\sigma = \{(a,x) \in A \times G : p(a) = r(x)\}$. Its groupoid structure is given by

(a,x) and (b,y) are composable iff x and y are ; then

$(a,x)(b,y) = (a(L(x)b)\sigma(x,y),xy)$

and $(a,x)^{-1} = ((L(x^{-1})a^{-1})\sigma(x^{-1},x)^{-1},x^{-1})$.

Define $i(a) = (a,p(a))$ and $j(a,x) = x$ and note that $k(x) = (r(x),x)$ is a covariant section. It is readily verified that (E_σ,i,j) is an extension and that its class depends only on the class of σ.

Conversely, if (E,i,j) is an extension of A by G and k is a covariant section, then σ defined by $i(\sigma(x,y)) = k(x)k(y)k(xy)^{-1}$ is a 2-cocycle in $Z^2(G,A)$. Its class is not affected by another choice of section or an equivalent extension. Finally ϕ: $(a,x) \to i(a)k(x) : E_\sigma \to E$ sets up an equivalence of E_σ and E.

The trivial extension is the semi-direct product of A and G.

Let us finally note that two similar groupoids have same cohomology groups with coefficients in a trivial constant module bundle. Explicitly, let $\phi : G \to H$ and $\psi : H \to G$ be two half-similarities ; $\phi\circ\psi \sim id_G$ and $\psi\circ\phi \sim id_H$. The maps $f \to f\circ\psi^n$: $C^n(G,A) \to C^n(H,A)$ and $g \to g\circ\phi^n$: $C^n(H,A) \to C^n(G,A)$ give isomorphisms of the cohomology groups.

A cohomology theory for inverse semi-groups may be given along the same lines. Suppose that $\mathcal{C}$ is some category. Let A_0 be a set. The set 2^{A_0} of all subsets of A_0, when ordered by inclusion, is a category : there is an arrow $V \to U$ precisely when $V \subset U$. A $\mathcal{C}$-sheaf $\mathcal{A}$ based on A^0 is a contravariant functor $U \to \mathcal{A}_U$ on 2^{A_0} to $\mathcal{C}$ (the morphism $\mathcal{A}_U \to \mathcal{A}_V$ corresponding to $V \subset U$ should thought of as the restriction morphism).

A partial isomorphism ϕ of $\mathcal{A}$ is a bijection $\phi : V \to U$, where V and U are subsets of A_0 together with isomorphisms $\phi : \mathcal{A}_{V'} \to \mathcal{A}_{\phi(V')}$, for any $V' \subset V$, compatible with the restriction morphisms, that is, such that for $V'' \subset V'$, the following diagram commutes

$$\begin{array}{ccc} \mathcal{A}_{V'} & \longrightarrow & \mathcal{A}_{\phi(V')} \\ \downarrow & & \downarrow \\ \mathcal{A}_{V''} & \longrightarrow & \mathcal{A}_{\phi(V'')} \end{array}.$$

Two partial isomorphisms ϕ and ϕ' may be composed : we have $\phi : V \to U$ and $\phi' : V' \to U'$; we let V'' be $\phi'^{-1}(U' \cap V)$ and U'' be $\phi(U' \cap V)$; $\phi'' = \phi \circ \phi'$ is the bijection $V'' \to U''$ obtained by composing ϕ and ϕ' ; and for $\underline{V} \subset V''$ we define $\phi'' : \mathcal{A}_{\underline{V}} \to \mathcal{A}_{\phi''(\underline{V})}$ by composing $\mathcal{A}_{\underline{V}} \xrightarrow{\phi'} \mathcal{A}_{\phi'(\underline{V})} \xrightarrow{\phi} \mathcal{A}_{\phi\circ\phi'(\underline{V})}$. The inverse of a partial isomorphism is defined in the obvious fashion. These operations make $\mathcal{J}so(\mathcal{A})$ = {partial isomorphisms of $\mathcal{A}$} into an inverse semi-group, that we call the isomorphism inverse semi-group of the $\mathcal{C}$-sheaf $\mathcal{A}$.

<u>1.15. Definition :</u> Let $\mathcal{G}$ be an inverse semi-group. A <u>$\mathcal{G}$-sheaf</u> $(\mathcal{A},\mathcal{L})$ is a $\mathcal{C}$-sheaf together with a homomorphism $\mathcal{L} : \mathcal{G} \to \mathcal{J}so(\mathcal{A})$ such that $\mathcal{L}^0 : \mathcal{G}^0 \to 2^{A_0}$ is an injection.

We let $\mathcal{G}^n$ be $\mathcal{G}\times\ldots\times\mathcal{G}$ n times for $n \geq 1$ and $\mathcal{G}^0$ be as before. Given a $\mathcal{G}$-sheaf $(\mathcal{A},\mathcal{L})$ of abelian groups, one can form the following cochain complex. A n-cochain is a function f from $\mathcal{G}^n$ to $\mathcal{A}$ which satisfies the conditions

(i) $f(s_0,s_1,\ldots,s_{n-1}) \in \mathcal{A}_{r(s_0 s_1 \ldots s_{n-1})}$;

(ii) f is compatible with the restriction maps, that is, if $U = r(s_0\, s_1 \ldots s_{n-1})$ and $V = r(t_0\, t_1 \ldots t_{n-1})$ where $t_i = e_i s$ for some idempotent element e_i then $f(t_0,t_1,\ldots,t_{n-1}) \in \mathcal{A}_V$ is the restriction of $f(s_0,s_1,\ldots,s_{n-1}) \in \mathcal{A}_U$ to V ; and

(iii) for $n > 0$, $f(s_0,\ldots,s_i,\ldots,s_{n-1}) \in 2^{A_0}$ whenever s_i is an idempotent element.

The set $C^n(\mathcal{G},\mathcal{A})$ of n-cochains is an abelian group under pointwise addition. The sequence

$$0 \to C^0(\mathcal{G},\mathcal{A}) \to C^1(\mathcal{G},\mathcal{A}) \to \ldots \to C^n(\mathcal{G},\mathcal{A}) \xrightarrow{\delta^n} C^{n+1}(\mathcal{G},\mathcal{A}) \to \ldots$$

where $\delta^0 f(s) = \mathcal{L}(s)\, f\circ d(s) - f\circ r(s)$

and
$$\delta^n f(s_0,\ldots,s_n) = \mathcal{L}(s_0)\, f(s_1,\ldots,s_n) + \sum_{i=1}^{n} (-1)^i f(s_0,\ldots,s_{i-1}s_i,\ldots,s_n) + (-1)^{n+1} f(s_0,\ldots,s_{n-1}),$$

is a cochain complex.

1.16. Definition : The group of n-cocycles and the group of n-coboundaries of this complex will be denoted respectively by $Z^n(\mathcal{G},\mathcal{A})$ and by $B^n(\mathcal{G},\mathcal{A})$. The n-th cohomology group $Z^n(\mathcal{G},\mathcal{A})/B^n(\mathcal{G},\mathcal{A})$ will be denoted $H^n(\mathcal{G},\mathcal{A})$.

Before giving the next definition, let us remark that $\mathcal{A} = \cup \mathcal{A}_U$, where U runs over $\mathcal{A}^0 = 2^{A_0}$, has a structure of inverse semi-group, where for $a \in \mathcal{A}_U$ and $b \in \mathcal{A}_V$, a + b is the element of $\mathcal{A}_{U\cap V}$ obtained by adding up the restrictions of a and b to $U\cap V$.

1.17. Definition : Let $(\mathcal{A}, \mathcal{L})$ be a $\mathcal{G}$-sheaf of abelian groups, noted multiplicatively. An extension of $\mathcal{A}$ by $\mathcal{G}$ is an exact sequence of inverse semi-groups

$$\mathcal{A}^0 \to \mathcal{A} \xrightarrow{i} \mathcal{E} \xrightarrow{j} \mathcal{G} \to \mathcal{G}^0 \quad \text{(we also write } (\mathcal{E},i,j))$$

compatible with the action of $\mathcal{G}$ on $\mathcal{A}$ in the sense that there exists a section k for j such that

(i) $k(e) = e$ for $e \in \mathcal{G}^0$ ($\mathcal{A}^0$, $\mathcal{E}^0$ and $\mathcal{G}^0$ are identified)

(ii) $k(s)\, i(a)\, k(s)^{-1} = i(\mathcal{L}(s)a)$ for $(a,s) \in \mathcal{A}\times\mathcal{G}$.

(iii) $k(es) = ek(s)$ and $k(se) = k(s)e$ for $e \in \mathcal{G}^0$, $s \in \mathcal{G}$.

Two extensions $(\mathcal{E},i,j)$ and $(\mathcal{E}',i',j')$ are equivalent if there exists an isomorphism $\phi : \mathcal{E} \to \mathcal{E}'$ such that $i' = \phi\circ i$ and $j = j'\circ\phi$. The set of equivalence classes of extensions with the Baer sum is an abelian group denoted Ext $(\mathcal{A},\mathcal{G})$ and just as before, Ext $(\mathcal{A},\mathcal{G})$ is isomorphic to $H^2(\mathcal{A},\mathcal{G})$.

1.18. Finally, we note the relationship between the cohomology of a groupoid G and the cohomology of the inverse semi-group of its G-sets, $\mathcal{G}$. Let (A,L) be a G-module bundle. One forms the following $\mathcal{G}$-sheaf of abelian groups based on A^0, $(\mathcal{A},\mathcal{L})$. For $U \subset A^0$, $\mathcal{A}_U$ = {sections of A defined on U} with its additive structure ; for $V\subset U$, the morphism $\mathcal{A}_U \to \mathcal{A}_V$ is the usual restriction map. The homomorphism $\mathcal{L} : \mathcal{G} \to \mathcal{J}so(\mathcal{A})$ is defined by : $\mathcal{L}(s)$ is the bijection $d(s) \to r(s)$ which sends u into $u \cdot s^{-1}$ and for $V \subset d(s)$ and $U = Vs^{-1} \subset r(s)$, $\mathcal{L}(s) : \mathcal{A}_V \to \mathcal{A}_U$ is given by $\mathcal{L}(s)\, h(u) = L(us)\, h(u \cdot s)$ for $h \in \mathcal{A}_V$. A cochain $f \in C^n(G,A)$ defines a cochain $\tilde{f} \in C^n(\mathcal{G},\mathcal{A})$. Namely $\tilde{f}(s_0,s_1,\ldots,s_{n-1})$ is the section of A defined on $r(s_0 s_1 \ldots s_{n-1})$ by $\tilde{f}(s_0,s_1,\ldots,s_{n-1})(u) = f(us_0,(u\cdot s_0)s_1,\ldots,(u\cdot s_0s_1\ldots s_{n-2})s_{n-1})$. It is compatible with the restriction maps. The map $f \mapsto \tilde{f}$ commutes with the coboundary operators, $\delta^n\tilde{f} = (\delta^n f)^{\sim}$. Therefore, if $f \in Z^n(G,A)$ (resp $B^n(G,A)$), then $\tilde{f} \in Z^n(\mathcal{G},\mathcal{A})$

(resp $B^n(\mathcal{G},\mathcal{A})$). Conversely, given $g \in C^n(\mathcal{G},\mathcal{A})$, we may define $f \in C^n(G,A)$ by $f(x_0,x_1,\ldots,x_{n-1}) = g(\{x_0\},\{x_1\},\ldots,\{x_{n-1}\})\ (r(x_0))$ where $\{x_0\},\{x_1\},\ldots,\{x_{n-1}\}$ are considered as G-sets. Then $g = \tilde{f}$. In conclusion $C^n(\mathcal{G},\mathcal{A}) \simeq C^n(G,A)$; $Z^n(\mathcal{G},\mathcal{A}) \simeq Z^n(G,A)$; $B^n(\mathcal{G},\mathcal{A}) \simeq B^n(G,A)$; $H^n(\mathcal{G},\mathcal{A}) \simeq H^n(G,A)$. We will use a topological version of this result in 2.14.

2. Locally Compact Groupoids and Haar Systems.

The definition of a topological groupoid and its immediate consequences can be found in [79], [26] page 23 and [68] page 26.

2.1. Definition : A topological groupoid consists of a groupoid G and a topology compatible with the groupoid structure :

(i) $x \mapsto x^{-1} : G \to G$ is continuous

(ii) $(x,y) \mapsto xy : G^2 \to G$ is continuous where G^2 has the induced topology from $G \times G$.

Consequences : $x \mapsto x^{-1}$ is a homeomorphism ; r and d are continuous ; if G is Hausdorff, G^0 is closed in G ; if G^0 is Hausdorff, G^2 is closed in $G \times G$. G^0 is both a subspace of G and a quotient of G (by the map r) ; the induced topology and the quotient topology coincide.

We will only consider topological groupoids whose topology is Hausdorff and, with the exception of section 4, locally compact. We will usually use Bourbaki's theory of integration on locally compact spaces [5,6,7].

If X is a locally compact space, $C_c(X)$ denotes the locally convex space of complex-valued continuous functions with compact support, endowed with the inductive limit topology.

2.2. Definition : Let G be a locally compact groupoid. A left Haar system for G consists of measures $\{\lambda^u, u \in G^0\}$ on G such that

(i) the support supp λ^u of the measure λ^u is G^u,

(ii) (continuity) for any $f \in C_c(G)$, $u \mapsto \lambda(f)(u) = \int f d\lambda^u$ is continuous, and

(iii) (left invariance) for any $x \in G$ and any $f \in C_c(G)$, $\int f(xy)\, d\lambda^{d(x)}(y) = \int f(y) d\lambda^{r(x)}(y)$.

This is Westman's definition ([77] p.2) of a left invariant continuous system of measures. It differs from Seda's definition ([68] p.27) in two respects : no measure on the unit space is given and continuity is required ; this last assumption is a rather severe restriction on the topology of G. In Section 4 of [68] and theorem 2 of [67], Seda gives conditions under which continuity holds automatically ; it seems preferable here to assume it as part of the definition.

The following results are easy consequences of the definition (cf. [77] 1.3 , 1.4).

2.3. Proposition : $\lambda: C_c(C) \to C_c(G^0)$ is a continuous sujection.

2.4. Proposition : Let G be a locally compact groupoid with a left Haar system. Then $r : G \to G^0$ is an open map, and the associated equivalence relation on the unit space is open.

2.5. Examples :

(a) A locally compact transformation group $G = U \times S$ has a distinguished left Haar system : $\lambda^u = \delta_u \times \lambda$, where δ_u is the point-mass at u and λ a left Haar measure for S.

(b) If G is a locally compact groupoid, then G^2 with the topology induced from $G \times G$ is also a locally compact groupoid. If $\{\lambda^u\}$ is a left Haar system for G, then $\{(\lambda^2)^x\}$ is a left Haar system for G^2 where

$$\int f\, d(\lambda^2)^x = \int f(x,z)\, d\lambda^{d(x)}(z) \text{ for } f \in C_c(G^2).$$

For example, if G is a group, $G^2 = G \times G$. As a groupoid, it is the groupoid associated with the transformation group (G,G) where G acts on itself by translation. Its left Haar system is $\delta_x \times \lambda$, where λ is a left Haar measure for G, as in example a.

(c) Let G be a locally compact principal groupoid. The map $d : G^u \to [u]$ is a bijection which gives to $[u]$ a locally compact topology, which can be different from the topology induced from G^0. An alternate definition for a left Haar system on G is :

a system of measures $\{\alpha_{[u]}, u \in G^0\}$ where

(i) $\alpha_{[u]}$ is a measure on $[u]$ of support $[u]$

(ii) for any $f \in C_c(G), u \mapsto \int f(u,v)d\alpha_{[u]}(v)$ is continuous (G is viewed as a subset of $G^0 \times G^0$).

These definitions are equivalent : if $\{\lambda_u\}$ is given, $\alpha_{[u]} = d_* \lambda^u$ depends only on $[u]$ and satisfies (i') and (ii') ; conversely if $\{\alpha_{[u]}\}$ is given, $\{\lambda^u\}$ is a left Haar system, where $\int f d\lambda^u = \int f(u,v)d\alpha_{[u]}(v)$.

(d) Let G be a locally compact group bundle, that is, a locally compact groupoid which is a group bundle in the sense of 1.2.d. Then a left Haar system, if it exists, is essentially unique in the sense that two left Haar systems $\{\lambda^u\}$ and $\{\nu^u\}$ differ by a continuous positive function h on G^0 : $\lambda^u = h(u)\nu^u$. The isotropy group bundle $G' = \{x \in G : d(x) = r(x)\}$ of a locally compact groupoid G is closed, hence locally compact. In the case where G is a transformation group, the existence of a left Haar system on G' is the assumption made in [37] (see beginning of the first section page 386) to determine the topological structure of the space of all irreducible induced representations of G.

(e) Let G be a locally compact group. The set S of subgroups of G becomes a compact Hausdorff space when equipped with Fell's topology [32]. $\underline{G} = \{(K,x) : K \in S, x \in K\} \subset S \times G$ with the topology induced from $S \times G$ and the groupoid structure : (K,x) and (L,y) are composable iff K = L, $(K,x)(K,y) = (K,xy)$, $(K,x)^{-1} = (K,x^{-1})$ is a locally compact group bundle, that we may call the subgroups bundle of G. It is shown in [32] that a left Haar system (λ^K) exists. It is essentially unique by d. For each $K \in S$, λ^K is a left Haar measure for K.

<u>2.6. Definition</u> : A locally compact groupoid is <u>r-discrete</u> if its unit space is an open subset.

<u>2.7. Lemma</u> : Let G be an r-discrete groupoid.

(i) For any $u \in G^0$, G^u and G_u are discrete spaces.

(ii) If a Haar system exists, it is essentially the counting measures system.

(iii) If a Haar system exists, r and d are local homeomorphisms.

Proof :

(i) An x in G_u^v defines a homeomorphism $y \mapsto xy : G^v \to G^u$; since {v} is open in G^v, {x} is open in G^u.

(ii) Let $\{\lambda^u\}$ be a left Haar system. Since G^u is discrete and λ^u has support G^u, every point in G^u has positive λ^u-measure. Let $g = \lambda(\chi_{G^0})$, where χ_{G^0} is the characteristic function of G^0. It is continuous and positive. Replacing λ^u by $g(u)^{-1}\lambda^u$, we may assume that $\lambda^u(\{u\}) = 1$ for any u. Then by invariance, $\lambda^v(\{x\}) = 1$ for any $x \in G_u^v$.

(iii) We assume, as we may, that λ^u is the counting measure on G^u. Let x be a point of G. A compact neighborhood V of x meets G^u in finitely many points x_i $i = 1,\dots,n$. If $x_i \neq x$, there exists a compact neighborhood V' of x contained in V, which does not contain x_i. Therefore, we may assume that $G^u \cap V = \{x\}$. Then $\lambda^{r(x)}(V)= 1$. By continuity of the Haar system, we may assume that $\lambda^u(V) = 1$ for any $u \in r(V)$. This shows that $r : V \to G^0$ is injective, hence a homeomorphisms onto r(V).

2.8. Proposition : For a locally compact groupoid G, the following properties are equivalent :

(i) G is r-discrete and admits a left Haar system,

(ii) $r : G \to G^0$ is a local homeomorphism,

(iii) the product map $G^2 \to G$ is a local homeomorphism, and

(iv) G has a base of open G-sets.

Proof :

(i) $\Rightarrow$ (ii) This has been shown in 7(iii).

(ii) $\Rightarrow$ (iii) If $(x,y) \in G^2$, we may choose a compact neighborhood U of x and a compact neighborhood V of y such that $r_{|V}$ and $d_{|V}$ are homeomorphisms onto their images ; $U \times V \cap G^2$ is then a compact neighborhood of (x,y) on which the product map is injective.

$$x'y' = x''y'' \Rightarrow r(x') = r(x'') \Rightarrow x' = x''$$

$$\text{and } d(y') = d(y'') \Rightarrow y' = y''.$$

(iii) $\Rightarrow$ (iv) If $x \in G$ and U is a neighborhood of x, we may find open sets V and W such that $x \in V \subset U$, $x^{-1} \in W \subset U^{-1}$ and the restriction of the product map to

$V \times W$ is injective. So $V \cap W^{-1}$ is the desired open G-set.

(iv) ⟹ (ii) Clear.

(iv) ⟹ (i) The groupoid G is r-discrete : for any $u \in G^0$, there is an open G-set s such that $u \in r(s) = ss^{-1} \subset G^0$ and by (iii) ss^{-1} is open in G. Let λ^u be the counting measure on G^u and f be in $C_c(G)$. Using a partition of the identity, one can write f as a finite sum of functions supported on open G-sets s. Therefore it is enough to consider a function f whose support is contained in an open G-set s. Then $\lambda(f)(u) = \lambda^u(f) = f(us)$: $\lambda(f)$ is continuous.

Q.E.D.

<u>2.9. Corollary :</u> A locally compact groupoid G is r-discrete and admits a left Haar system iff G^2 is r^2-discrete and admits a left Haar system.

<u>2.10. Definition :</u> Let G be an r-discrete groupoid. Its <u>ample semi-group</u> $\mathcal{G}$ is the semi-group of its compact open G-sets.

This terminology, introduced by W. Krieger in [52], will be justified at the end of the section. The case of interest is when G admits a cover of compact open G-sets. Then G has a base of open G-sets, with sub-base $\{Us : U$ open subset of G^0 and $s \in \mathcal{G}\}$, therefore G admits a left Haar system. We do not know if there exist r-discrete groupoids which have a Haar system but do not have a cover of compact open G-sets. If G has a cover of compact open G-sets, it is completely described by $(G^0, \mathcal{G})$ in the sense that its groupoid structure as well as its topology may be recovered from G^0, $\mathcal{G}$ and the map r. If $x \in s$, with $s \in \mathcal{G}$, x^{-1} is defined by $s^{-1}\{r(x)\} = x^{-1}$. If $x \in s$, $y \in t$ with $s, t \in \mathcal{G}$ and $d(x) = r(y)$, xy is defined by $\{xy\} = \{r(x)\}st$. We have just seen that $\{Us : U$ open subset of G^0, $s \in \mathcal{G}\}$ is a sub-base for the topology of G.

Let us describe next the r-discrete principal groupoids which admit a cover of compact open G-sets.

<u>2.11. Definition :</u> Let U be a locally compact space and s a partial homeomorphism of U, defined on a compact open subset r(s) onto a compact open subset d(s). Let us say that s is <u>relatively free</u> if its set of fixed points $\{u \in r(s) : u \cdot s = u\}$ is (compact and) open. Let us say that an inverse semi-group $\mathcal{G}$ of partial homeomorphisms

defined on compact open subsets of U acts relatively freely if each $s \in \mathcal{G}$ is relatively free.

2.12. Definition : Let U be a locally compact space and $\mathcal{G}$ an inverse semi-group of partial homeomorphisms defined on compact open subsets of U. Let us say that $\mathcal{G}$ is ample if

(i) for any compact open set e in U, the identity map id_e belongs to $\mathcal{G}$.

(ii) for any finite family (s_i) $i=1,\dots,n$ in $\mathcal{G}$ such that $r(s_i) \cap r(s_j) = \emptyset$ and $d(s_i) \cap d(s_j) = \emptyset$ for $i \neq j$, there exists s in $\mathcal{G}$ denoted by $\sum s_i$ such that $u \cdot s = u \cdot s_i$ for $u \in r(s_i)$.

2.13. Proposition : Let U be a locally compact space and $\mathcal{G}$ an inverse semi-group of partial homeomorphisms defined on compact open subsets of U. Let G be the principal groupoid associated with the equivalence relation

$u \sim v$ iff there exists $s \in \mathcal{G} : u = v \cdot s$

Then the following properties are equivalent.

(i) G has a structure of r-discrete groupoid with a cover of compact open G-sets such that U becomes its unit space and its ample semi-group is the ample inverse semi-group generated by $\mathcal{G}$.

(ii) $\mathcal{G}$ acts relatively freely on U.

Proof :

(i) ⟹ (ii) Let s and t be two compact open G-sets of G. Then $s \cap t$ is a compact open G-set of G. Thus, if $s \in \mathcal{G}$

$\{u \in r(s) : u \cdot s = u\} = s \cap r(s)$ is compact open in $G^0 = V$.

(ii) ⟹ (i) For $s \in \mathcal{G}$, let $s = \{(u,us) : u \in r(s)\}$. We define on G the topology which has as sub-base $\{Vs : V \text{ open in } U \text{ and } s \in \mathcal{G}\}$. It makes G into a r-discrete groupoid admitting a cover of compact-open sets, namely $\mathcal{G}$. The induced topology on $G^0=U$ is identical to the original one. Finally, let s be a compact open G-set. It may be covered by finitely many open G-sets in $\mathcal{G}$. Hence there exists a finite family (s_i) $i = 1,\dots,n$ in $\mathcal{G}$ and a finite family (U_i) $i = 1,\dots,n$ of compact open sets of U such that $U_i \cap U_j = \emptyset$, $U_i \cdot s_i \cap U_j \cdot s_j = \emptyset$ for $i \neq j$ and $s = \bigcup_{i=1}^{n} U_i s_i$.

Q.E.D.

2.14. In the topological setting, we make the following adjustments to the cohomology theory given in the first section (see [79] p.24).

(a) In 1.11, we require that A be a locally compact group bundle and we require that for any continuous section $u \mapsto a_u$ of $p : A \to A_0$, the function $x \mapsto L(x)a_{d(x)}$ should be continuous.

(b) We give to G^n the topology induced from the product topology on $G\times\ldots\times G$ n-times and consider continuous cochains only. It will be implicit that $Z^n(G,A)$, $B^n(G,A)$ and $H^n(G,A)$ refer to the continuous cohomology.

If G is an r-discrete groupoid which admits a cover of compact open G-sets, the results of 1.18 are still valid when $\mathcal{G}$ is interpreted as the ample semi-group of G. Given $g \in C^n(\mathcal{G},\mathcal{A})$ (notations of 1.18), we define $f \in C^n(G,A)$ by $f(x_0,x_1,\ldots,x_{n-1}) = g(s_0,s_1,\ldots,s_{n-1})(r(x_0))$ where $s_0,s_1,\ldots,s_{n-1} \in \mathcal{G}$ and $x_0 \in s_0$, $x_1 \in s_1,\ldots,x_{n-1} \in s_{n-1}$. By assumption, there exist $s_0, s_1,\ldots,s_{n-1}$ with these properties. Moreover, the condition that g be compatible with the restriction maps shows that f is well defined. Finally f is continuous since its restriction to $s_0 \times s_1 \times \ldots \times s_{n-1}$ is continuous. Thus $H^n(G,A) \approx H^n(\mathcal{G},\mathcal{A})$.

3. Quasi-Invariant Measures

Let G be a locally compact groupoid with left Haar system $\{\lambda^u\}$. Let $\lambda_u = (\lambda^u)^{-1}$ be the image of λ^u by the inverse map $x \to x^{-1}$. Then $\{\lambda_u\}$ is a right Haar system.

3.1. Definition : Let μ be a measure on G^0. The measure on G induced by μ is $\nu = \int\lambda^u d\mu(u)$. The measure on G^2 induced by μ is $\nu^2 = \int\lambda_u\times\lambda^u \, d\mu(u)$. The image of ν by the inverse map is $\nu^{-1} = \int\lambda_u d\mu(u)$.

These measures are well defined since the system $\{\lambda^u\}$ of measures on G and the system $\{\lambda_u\times\lambda^u\}$ of measures on G^2 are μ-adequate (Bourbaki [6] 3.1) ; ν^2 is also the measure on G^2 induced by ν^{-1} with respect to the Haar system 2.5.b.

3.2. Definition : A measure μ on G^0 is said to be quasi-invariant if its induced measure ν is equivalent to its inverse ν^{-1}. A measure belonging to the class of μ is also quasi-invariant ; we say that the class is invariant.

If G is second countable and μ is a quasi-invariant measure on G^0, then (G,C), where C is the class of ν, is a measure groupoid in the sense of P.Hahn [44] p.15 and (ν,μ) is a Haar measure for (G,C) (definition 3.11 p. 39). Most of the results and techniques of this section can be found in [44] and in [61].

The cohomology theory for measure groupoids is developed in [76] ; the discrete principal case is studied thoroughly in [31]. The relevant fact here is that to each quasi-invariant measure is associated a 1-cocycle with values in $\mathbb{R}^+_*$, whose class depends on the measure class only.

3.3.Proposition : Let μ be a quasi-invariant measure on G^0 and D a locally ν-integrable positive function such that $\nu = D\nu^{-1}$, then

(i) for ν^2 a.e. (x,y), $D(xy) = D(x)D(y)$ and for ν a.e. x, $D(x^{-1}) = D(x)^{-1}$;

(ii) if $\mu' = g\mu$ where g is a locally μ-integrable positive function, $D' = (g \circ r)D\,(g \circ d)^{-1}$ satisfies $\nu' = D'\nu'^{-1}$.

Proof :

(i) (see also [44], theorem 3.1, p. 31) One shows that $D^2(x,y) = D(y)$ and $\underline{D}^2(x,y) = D(xy)D(x)^{-1}$ are versions of the Radon-Nikodym derivative $\dfrac{d\nu^2}{(d\nu^2)^{-1}}$; this gives the first assertion.

(ii) Straightforward.

Q.E.D.

This proposition shows that the Radon-Nikodym derivative of ν with respect to ν^{-1} (defined ν a.e.) is a one-cocycle with values in $\mathbb{R}^*_+$ in the sense of [76] §3 and that its class depends on the class of μ only.

3.4. Definition : Let μ be a quasi-invariant measure on G^0 ; (a version of) the Radon-Nikodym derivative $D = \dfrac{d\nu}{d\nu^{-1}}$ is called the modular function (or the Radon-

Nikodym derivative) of μ.

If G is a group, the point mass at e is, up to a scalar multiple, the only quasi-invariant measure on $G^0 = \{e\}$. Its modular function in the sense of 3.4 equals a.e. the modular function of the group.

It will be convenient for later purpose to choose a particular symmetric measure in the class of ν, where symmetric means equal to its inverse (the inverse of a measure on G is its image under the inverse map). We choose $\nu_0 = D^{-1/2}\nu$ and call it the symmetric measure induced by μ.

3.5. Definition : Let μ be a quasi-invariant measure on G^0. A measurable set A in G^0 is almost invariant (with respect to μ) if for ν a.e. $x, r(x) \in A$ iff $d(x) \in A$. The measure μ is called ergodic if every almost invariant measurable set is null or conull.

Let X and Y be locally compact spaces and p a continuous map from X onto Y. If X is σ-compact, it is possible to define the image $p_* C$ of a measure class C on X : one chooses a probability measure μ in the class of C and defines $p_* C$ as the class of $p_* \mu$, where $p_* \mu(E) = \mu(p^{-1}(E))$; $p_* C$ depends only on the class of C. As it is easier to deal with measures rather than with measure classes, one introduces the notion of pseudo-image of a measure (see [6]) : a pseudo-image of an arbitrary measure μ on X is a measure in the image $p_* C$ of the class C of μ.

3.6. Proposition : Let μ be a measure on G^0 and $[\mu]$ be a pseudo-image by d of the induced measure ν. Then

(i) $[\mu]$ is a quasi-invariant ; and

(ii) μ is quasi-invariant iff $\mu \sim [\mu]$.

Proof :

(i) Let $[\nu] = \int \lambda^v \, d[\mu](v)$ and f be a non-negative measurable function. Then $[\nu](f) = 0$ iff for $[\mu]$ a.e. v, $\lambda^v(f) = 0$;

iff for ν a.e. x and $\lambda^{d(x)}$ a.e. y, $f(y) = 0$;

iff for μ a.e. u, λ^u a.e. x and $\lambda^{d(x)}$ a.e. u, $f(y) = 0$

iff for μ a.e. u, λ^u a.e. x and $x^{-1}\lambda^u$ a.e. y, $f(y) = 0$;

iff for μ a.e. u, λ^u a.e. x and λ^u a.e. z, $f(x^{-1}z) = 0$;

iff for μ a.e. u, λ^u a.e. x and λ^u a.e. z, $f(z^{-1}x) = 0$,

by Fubini's theorem ;

iff for μ a.e. u, λ^u a.e. x and $\lambda^{d(x)}$ a.e. y, $f(y^{-1}) = 0$;

iff for ν a.e. x and $\lambda^{d(x)}$ a.e. y, $f(y^{-1}) = 0$;

iff $[\nu]^{-1}(f) = 0$.

(ii) If μ is quasi-invariant, μ is a pseudo-image by d of $\nu^{-1} \sim \nu$.

3.7. Definition : Let μ be a measure on G^0. Then a measure $[\mu]$ as above is called a saturation of μ.

3.8. Proposition : Let α_u be the saturation of the point mass at u, that is, a pseudo-image of λ^u. Then

(i) the class of α_u depends only on the orbit $[u]$;

(ii) α_u is ergodic ; and

(iii) every quasi-invariant measure carried by $[u]$ is equivalent to α_u.

Proof :

(i) Let N be a subset of $[u]$ and v be in $[u]$. Since $\lambda^u = x\,\lambda^v$ for $x \in G^u_v$, $d^{-1}(N)$ is λ^u-negligible iff it is λ^v-negligible.

(ii) The ergodicity of a transitive quasi-invariant measure is well known (e.g. [61], theorem 4.6, p. 278). Suppose that A is almost invariant and has positive measure and let ν be $\int \lambda^v \, d\alpha_u(v)$.

Then $0 = \nu[d^{-1}(G^0 \setminus A) \cap r^{-1}(A)] = \int_A \lambda^v[d^{-1}(G^0 \setminus A)]d\alpha_u(v)$. Hence, for some v in A, $\lambda^v[d^{-1}(G^0 \setminus A)] = 0$ and by (i) $\alpha_u(G^0 \setminus A) = 0$.

(iii) (cf. [61], lemma 4.5, p. 277). Let μ be a quasi-invariant probability measure such that $\mu([u]) = 1$ and let ν be its induced measure on G. Then α_u is a pseudo-image of ν by d :

$\nu[d^{-1}(A)] = 0$ iff for μ a.e. v, $\lambda^v[d^{-1}(A)] = 0$;

iff for μ a.e. v, $\lambda^u[d^{-1}(A)] = 0$ because of (i) ;

iff $\alpha_u(A) = 0$.

But so is μ because of quasi-invariance :

$$\nu[d^{-1}(A)] = 0 \text{ iff } \nu^{-1}[d^{-1}(A)] = 0 \;;$$
$$\text{iff for } \mu \text{ a.e. } v,\ \lambda_v\ [d^{-1}(A)] = 0 \;;$$
$$\text{iff } \mu(A) = 0.$$

Q.E.D.

If G is the groupoid of a transitive transformation group (U,S), the class of α_u is the unique invariant measure class on U. This case is well known (e.g. [74], theorem 8.19, p.25).

3.9. Definition : A transitive measure is a quasi-invariant measure carried by an orbit. Up to equivalence, there exists one and only one transitive measure on the orbit [u]; it will be denoted $\alpha_{[u]}$. A quasi-invariant ergodic measure which is not transitive is called properly ergodic. A quasi-orbit is an equivalence class of quasi-invariant ergodic measures.

3.10. Proposition : Suppose that G is second countable. The modular function D of the transitive measure $\alpha_{[u]}$ can be chosen such that $D_{|G(v)}$ = modular function of G(v) for $\alpha_{[u]}$ a.e.v.

Proof : This is in theorem 4.4 p. 48 of [44]. An alternate proof is to use a similarity between the essentially transitive groupoid $(G,\alpha_{[u]})$ and the group G(u) (cf. [69], theorem 6.19).

3.11. A well known theorem of J. Glimm [36] states that, for a second contable locally compact transformation group G, the following properties are equivalent :

(i) every orbit is locally closed ;

(ii) the orbit space G^0/G with the quotient topology is T_0 ;

(iii) every quasi-orbit is transitive.

We do not know if this can be generalized to arbitrary second countable locally compact groupoids with Haar system. The implications (i) $\Longrightarrow$(ii) $\Longrightarrow$(iii) may be obtained as in [36].

3.12. Definition : An invariant measure is a quasi-invariant measure whose modular function is equal to 1.

3.13. Definition : ([55] p. 448). Suppose G principal. A quasi-orbit is called

(i) type I if it is transitive,

(ii) type II_1 if it is properly ergodic and contains a finite invariant measure,

(iii) type II_∞ if it is properly ergodic and contains an infinite invariant measure, and

(iv) type III if it is properly ergodic and contains no invariant measure.

3.14. Definition : A principal groupoid is of type I if it has type I quasi-orbits only.

The notion of invariant measure can be extended as follows. Before giving the definition, recall that $Z^1(G, \mathbb{R})$ is the group of continuous homomorphisms of G into $\mathbb{R}$. Let c be in $Z^1(G, \mathbb{R})$, then we denote by Min(c) the set of u's in G^0 such that $c(G_u)$ is in $[0,\infty)$ and by Max(c) the set Min (-c).

3.15. Definition : Let $c \in Z^1(G, \mathbb{R})$ and $\beta \in [-\infty, +\infty]$. We say that a measure μ on G^0 satisfies the (c,β) KMS condition if

(i) when β is finite, μ is quasi-invariant and its modular function D is equal to $e^{-\beta c}$;

(ii) when $\beta = \pm\infty$, the support of μ is contained in Min (± c). A (c,∞) KMS probability measure is also called a ground state for c. The point mass at u is called a physical ground state if $\mathrm{Min}(c) \cap [u] = \{u\}$.

The terminology will be justified in the section 4 of the second chapter. However the condition $D = e^{-\beta c}$ is closer to the classical Gibbs Ansatz for equilibrium states than to the analytic form of the KMS conditions (cf. example 3.1.6).

3.16. Proposition :

(i) Note first that $c^{-1}(0)$ is a locally compact groupoid. If G is r-discrete with Haar system, then so is $c^{-1}(0)$.

(ii) Suppose that G is r-discrete and that β is finite. Then, a (c,β) KMS measure for G is an invariant measure for $c^{-1}(0)$.

(iii) The subset Min(c) is closed in G^0.

(iv) The subset Min(c) is invariant under $c^{-1}(0)$, that is, if $x \in c^{-1}(0)$ and $d(x) \in \mathrm{Min}(c)$, then $r(x) \in \mathrm{Min}(c)$.

(v) The reduction of G to Min(c) is equal to the reduction of $c^{-1}(0)$ to Min(c).

Proof : Assertions (i) and (ii) are clear.

(iii) If $u \notin \mathrm{Min}(c)$, there exists $x \in G$ such that $d(x) = u$ and $c(x) < 0$. Let V be an open neighborhood of c such that $c(y) < 0$ for $y \in V$. Then d(v) is an open neighborhood of u and $d(V) \cap \mathrm{Min}(c) = \phi$.

(iv) Let $x \in c^{-1}(0)$ with $d(x) \in \mathrm{Min}(c)$. For any $y \in G_{r(x)}$, $yx \in G_{d(x)}$ and $c(y) = c(y) + c(x) = c(yx) \geq 0$. This shows that $r(x) \in \mathrm{Min}(c)$.

(v) If $d(x) \in \mathrm{Min}(c)$, $c(x) \geq 0$ and if $r(x) \in \mathrm{Min}(c)$, $-c(x) = c(x) \geq 0$, hence $c(x) = 0$.

Q.E.D.

3.17. Proposition : (cf. [65], theorem 7.5, page 26) A limit point (with respect to the vague convergence of measures) of (c,β) KMS measures when $\beta \to \infty$ is a (c,∞) KMS measure.

Proof : Suppose that μ_β tends to μ as β tends to ∞ and suppose that the modular function of μ_β is $e^{-\beta c}$. Let ν_β be the induced measure and let ν be the induced measure of μ. Then ν_β tends to ν and ν_β^{-1} tends to ν^{-1} as β tends to ∞. Therefore, for every non-negative f in $C_c(G)$,

$$\begin{aligned} \int fc\,d\nu^{-1} &= \lim \int fc\,d\nu_\beta^{-1} \\ &= \lim \int fce^{\beta c}\,d\nu_\beta \\ &= \lim \left(\int_{c \geq 0} fce^{\beta c}\,d\nu_\beta + \int_{c < 0} fce^{\beta c}\,d\nu_\beta\right). \end{aligned}$$

Since $ce^{\beta c}$ tends to 0 uniformly on $c < 0$, the second integral tends to 0. Hence $\int fc\,d\nu^{-1}$ is non-negative for every non-negative $f \in C_c(G)$.

Thus, c is non-negative on the support of ν^{-1}, which is $d^{-1}(\mathrm{supp}\mu)$. That is, suppμ is contained in Min(c).

Q.E.D.

The last part of this section is devoted to the study of the relationship between the notion of quasi-invariance given in 3.2. and the usual notion of quasi-invariance under an inverse semi-group of transformations.

Let us first look at the case of a transformation group (U,S). Let $G = U \times S$ be the associated groupoid. The measure on G induced by the measure μ on U is $\nu = \mu \times \lambda$, where λ is a left Haar measure of S. With respect to the groupoid G, the group S acts in two different ways :

(i) The horizontal action is the action of S on U. One says that μ is quasi-invariant if it is quasi-invariant under this action, that is, $\mu \sim \mu \cdot s$ for any $s \in S$.

(ii) The vertical action is the action of S on itself, or rather on each fiber $\{u\} \times S$. One notes that λ is quasi-invariant under this action. If we let S act on the right, $\frac{d\lambda \cdot s^{-1}}{d\lambda}$ is equal to $\delta(s)$, where δ is the modular function of S.

Before studying the general case, let us establish some conventions : Let (X,μ) and (Y,ν) be two measure spaces and $s : X \to Y$ a bimeasurable bijection from X onto Y. The image of x by s is written $x \cdot s$ and the image of μ by s is written $\mu \cdot s$. Thus, $\int f(y)d(\mu \cdot s)(y) = \int f(x \cdot s)d\mu(x)$ for $f \in C_c(Y)$. If $\mu \cdot s$ is absolutely continuous with respect to ν, $\frac{d\mu \cdot s}{d\nu}$ denotes the Radon-Nikodym derivative of $\mu \cdot s$ with respect to ν. One says that s is non-singular if it induces an isomorphism of the measure algebras.

<u>3.18. Definition :</u> Let G be a locally compact groupoid with Haar system $\{\lambda^u\}$. Let μ be a measure on G^0, not necessarily quasi-invariant, and ν be its induced measure. Let s be a G-set measurable with respect to the completion of ν.

(i) We say that ν is <u>quasi-invariant under s</u> (or s is non-singular with respect to ν) if the map from $(d^{-1}[d(s)], \nu_{|d^{-1}[d(s)]})$ to $(d^{-1}[r(s)], \nu_{|d^{-1}[r(s)]})$ defined by the rule $x \mapsto xs^{-1}$ is non singular.
The Radon-Nikodym derivative $\frac{d\nu s^{-1}}{d\nu}$ (where we write ν instead of the appropriate restriction) will be denoted by $\delta(\cdot,s)$ and called the <u>vertical Radon-Nikodym derivative</u> of s (with respect to ν).

(ii) We say that <u>μ is quasi-invariant under s</u> if the map from $(d(s), \mu_{|d(s)})$ to $(r(s), \mu_{|r(s)})$ defined by the rule $u \mapsto u \cdot s^{-1}$ is non-singular. The Radon-Nikodym derivative $\frac{d\mu \cdot s^{-1}}{d\mu}$ will be denoted by $\Delta(\cdot,s)$ called the <u>horizontal Radon-Nikodym derivative</u> of s (with respect to μ).

Remark : Since we assume that G is second countable, (G,ν) is a standard measure space. Therefore, if s is a measurable G-set, r(s) is measurable and the map from r(s) to s sending u to us is measurable.

3.19. Proposition : With the notations of the previous definition, assume that μ is quasi-invariant. Then the vertical Radon-Nikodym derivative of a non-singular measurable G-set s with respect to ν depends on d(x) only. More precisely, there exists a function $u \mapsto \delta(u,s)$ defined on r(s), positive and measurable and which we still call the vertical Radon-Nikodym derivative of s, such that

$$\delta(d(x), s) = \frac{d\nu s^{-1}}{d\nu}(x) \text{ for } \nu \text{ a.e. } x \text{ in } d^{-1}[r(s)].$$

Proof : Let $\underline{\delta}(x) = \frac{d(\nu s^{-1})}{d\nu}(x)$ be the vertical Radon-Nikodym derivative of s with respect to ν. Since $\nu s^{-1} = \int_{d(s)} (\lambda^u s^{-1})\, d\mu(u)$ and $\nu s^{-1} = \int_{d(s)} \mu(\underline{\delta}\lambda^u)\, d\mu(u)$ are two r-decompositions of νs^{-1}, there exists a μ-conull set U in G^0 such that for every u in U, $\lambda^u s^{-1} = \underline{\delta}\lambda^u$. That is, for u in U and λ^u a.e. x in $d^{-1}[r(s)]$, $\underline{\delta}(x) = \frac{d(\lambda^u s^{-1})}{d\lambda^u}(x)$. The commutativity of left and right multiplication allows us to write, for any x in G_U and any positive measurable f,

$$\begin{aligned}
\int f(y)\underline{\delta}(xy)\, d\lambda^{d(x)}(y) &= \int f(x^{-1}y)\, \underline{\delta}(y)\, d\lambda^{r(x)}(y) \\
&= \int f(x^{-1}ys^{-1})\, d\lambda^{r(x)}(y) \\
&= \int f(ys^{-1})\, d\lambda^{d(x)}(y) \\
&= \int f(y)\, \underline{\delta}(y)\, d\lambda^{d(x)}(y) .
\end{aligned}$$

Hence, for any x in G_U and $\lambda^{d(x)}$ a.e. y, $\underline{\delta}(xy) = \underline{\delta}(y)$. Therefore, if ϕ is a positive measurable function such that $\int \phi\, d\lambda_u = 1$ for u in U, the function δ defined in r(s) by

$$\delta(u) = \int \underline{\delta}(x)\ \phi(x)\, d\lambda_u(x)$$

has the required property. Indeed, since U is μ-conull, $d^{-1}(U)$ is ν^{-1}-conull and $r^{-1}(U)$ is ν-conull and since μ is quasi-invariant, $G_U = d^{-1}(U) \cap r^{-1}(U)$ is ν^{-1}-conull, hence λ_u-conull for μ a.e. u. Thus, for μ a.e. u and any positive measurable f,

$$\begin{aligned}
\int f(y)\ \delta \circ d(y)\, d\lambda^u(y) &= \int f(y)\ \underline{\delta}(x)\ \phi(x)\, d\lambda_{d(y)}(x)\, d\lambda^u(y), \\
&= \int f(y)\ \underline{\delta}(xy)\ \ \phi(xy)\, d\lambda_u(x)\, d\lambda^u(y), \\
&= \int(\int f(y)\ \ \underline{\delta}(xy)\ \ \phi(xy)\, d\lambda^u(y))\, d\lambda_u(x), \\
&= \int f(y)\ \underline{\delta}(y)\ (\int \phi(xy) d\lambda_u(x))\, d\lambda^u(y), \\
&= \int f(y)\ \ \underline{\delta}(y)\, d\lambda^u(y)\ ; \text{ therefore}
\end{aligned}$$

$$\int f(y)\ \delta\circ d(y)\ d\nu(y) = \int f(y)\ \underline{\delta}(y)\ d\nu(y),$$
$$= \int f(ys^{-1})\ d\nu(y).$$

Q.E.D.

3.20. Proposition : Let μ be a quasi-invariant measure, ν its induced measure and s a measurable G-set. Then the following properties are equivalent :

(i) ν is quasi-invariant under s.

(ii) μ is quasi-invariant under s. Moreover, if these conditions are satisfied, the vertical and the horizontal Radon-Nikodym derivatives of s with respect to μ, $\delta(\cdot,s)$ and $\Delta(\cdot,s)$, are related by the equation $\delta(u,s) = D(us)\ \ \Delta(u,s)$ for μ a.e. u in r(s), where D is the modular function of μ.

Proof : Suppose that (i) holds. Given a non-negative measurable function h defined on r(s), there exists a non-negative measurable function h defined on $d^{-1}[r(s)]$ such that $h(u) = \int f(x)\ d\lambda_u(x)$ for $u \in r(s)$(cf 2.3). Then,

$$\int h(u\cdot s^{-1})\ d\mu(u) = \int f(x)\ d\lambda_{u\cdot s^{-1}}(x)\ d\mu(u)$$
$$= \int f(xs^{-1})\ d\lambda_u(x)\ d\mu(u) \text{(by right invariance of } \{\lambda_u\})$$
$$= \int f(xs^{-1})\ D^{-1}(x)\ d\nu(x)$$
$$= \int f(x)\ D^{-1}(xs)\ \delta(d(x),s)\ d\nu(x)$$
$$= \int f(x)\ D^{-1}(d(x)s)\ \delta(d(x),s)\ D^{-1}(x)\ d\nu(x)$$
$$= \int f(x)\ D^{-1}(us)\ \delta(u,s)\ d\lambda_u(x)\ d\mu(u)$$
$$= \int h(u)\ D^{-1}(us)\ \delta(u,s)\ d\mu(u).$$

Hence μ is quasi-invariant under s and $\frac{d(\mu s^{-1})}{d\mu}(u) = D^{-1}(us)\delta(u,s)$ for μ a.e. u in r(s).

Conversely, suppose that (ii) holds. Then, for any non-negative measurable function f defined on $d^{-1}[r(s)]$,

$$\int f(xs^{-1})\ d\nu(x) = \int f(xs^{-1})\ D(x)\ d\lambda_u(x)\ d\mu(u)$$
$$= \int f(x)\ D(xs)\ d\lambda_{us^{-1}}(x)\ d\mu(u) \text{(by right invariance of } \{\lambda_u\})$$
$$= \int f(x)\ D(xs)\ d\lambda_u(x)\ \Delta(u,s)d\mu(u)$$
$$= \int f(x)\ D(d(x)s)\ \Delta(d(x),s)\ D(x)\ d\nu^{-1}(x)$$
$$= \int f(x)\ D(d(x)s)\ \Delta(d(x),s)\ d\nu(x).$$

This shows that ν is quasi-invariant under s and that

$$\frac{d\nu s^{-1}}{d\nu}(x) = D(d(x)s)\,\Delta(d(x),s) \text{ for } \nu \text{ a.e. } x \text{ in } d^{-1}[r(s)].$$

Q.E.D.

3.21. Case of a transformation group.

Let us look back to the case of a transformation group (U,S). With above notation, $G = U \times S$ and $\lambda^u = \delta_u \times \lambda$ where λ is a left Haar measure of S. For any measure μ on U and any G-set $s = \{u, u \cdot s) : u \in U\}$ where s is an element of S, the induced measure $\nu = \mu \times \lambda$ is quasi-invariant under s and $\delta(u,s) = \delta(s)$ where δ is the modular function of S. It is known (e.g. [61], theorem 4.3, page 276) that μ is quasi-invariant in the sense of 3.2 iff it is quasi-invariant under the group S. The horizontal Radon-Nikodym derivative $\Delta(u,s)$ is the usual Radon-Nikodym cocycle of the action. If μ is quasi-invariant, it follows from 3.20 that its modular function is

$$D(u,s) = \delta(s)/\Delta(u,s).$$

3.22. Case of an r-discrete groupoid.

Since the counting measure λ^u is invariant under any G-set s, the vertical Radon-Nikodym $\delta(u,s)$ is identically equal to 1, independently of any measure μ on G^0. Suppose that G admits a cover of compact open G-sets and let $\mathcal{G}$ be its ample semi-group (definition 2.10). Then a measure μ on G^0 is quasi-invariant iff it is quasi-invariant under $\mathcal{G}$. Indeed if μ is quasi-invariant, by 3.20 any compact open G-set leaves μ quasi-invariant. Conversely, if μ is quasi-invariant under $\mathcal{G}$, it is quasi-invariant since any compact set can be covered by finitely many compact open G-sets.

3.23. Case of a principal and transitive groupoid.

Let X be a locally compact space. As in 1.2.c, the graph $X \times X$ of the transitive equivalence relation on X (that is, any two elements of X are equivalent) has a structure of groupoid. With the product topology, it is a locally compact groupoid.

As in 2.5.c, any measure α on X with support equal to X defines a Haar system on $X \times X$. The transitive measure X induces the product measure $\alpha \times \alpha$. A measurable G-set s is non-singular

with respect to $\alpha \times \alpha$ iff it is the graph of a non-singular transformation of (X,α). The horizontal and vertical Radon-Nikodym derivatives of s with respect to α

are equal :

$$\Delta(x,s) = \delta(x,s) = \frac{d\alpha s^{-1}}{d\alpha}(x) \text{ for } \alpha \text{ a.e. } x \text{ in } r(s)$$

The measure α is invariant because its modular function is identically equal to 1.

We have defined in 3.18 the notion of a non-singular measurable G-set with respect to the induced measure ν of a measure μ on G^0. It will be useful to have a definition depending only on the groupoid G an the Haar system $\{\lambda^u\}$.

<u>3.24. Definition :</u> Let G be a locally compact groupoid.

(i) A G-set s will be called a <u>Borel G-set</u> [resp. a <u>continuous G-set</u>] if the restriction of each of the maps r and d to s is a Borel isomorphism onto a Borel subset of G^0 [resp. a homeomorphism onto an open subset of G^0].

(ii) Suppose that G has a Haar system $\{\lambda^u\}$. <u>A non-singular Borel G-set</u> [resp. <u>non-singular continuous G-set</u>] is a Borel G-set [resp. a continuous G-set] such that there exists a Borel [resp. continuous] positive function on r(s) bounded above and below on compact sets, denoted $\delta(\cdot,s)$ and called the vertical Radon-Nikodym derivative of s, such that

$$\delta(d(x),s) = \frac{d\lambda^u s^{-1}}{d\lambda^u}(x) \text{ for every } u \in G^0 \text{ and } \lambda^u \text{ a.e. } x \in d^{-1}[r(s)].$$

Thus, a non-singular Borel G-set s is non singular with respect to the induced measure ν of every measure μ on G^0 and

$$\delta(d(x),s) = \frac{d\nu s^{-1}}{d\nu}(x) \text{ for } \nu \text{ a.e. } x \in d^{-1}[r(s)].$$

<u>3.25. Examples :</u> In the case of a transformation group (U,S), the G-set s = $\{(u,u\cdot s) : u \in V\}$ where V is an open subset of U and $s \in S$, is a non-singular continuous G-set. Its vertical Radon-Nikodym derivative is $\delta(u,s) = \delta(s)$ for $u \in V$, where $\delta(s)$ the modular function of S evaluated at s. In the case of a r-discrete groupoid, any open G-set s is a non-singular continuous G-set. We have already observed that its vertical Radon-Nikodym derivative $\delta(u,s)$ is equal to 1, for $u \in r(s)$.

<u>3.26.</u> The set of non-singular Borel G-sets [resp.non-singular continuous G-sets] is an inverse semi-group under the operations $(s,t) \to st$ and $s \to s^{-1}$. We call it the Borel ample semi-group of G and denote it $\mathcal{G}_b$ [resp. the continuous ample semi-group of G and write $\mathcal{G}_c$]. Let us note the following formulas : for $s,t \in \mathcal{G}_b$,

$\delta(u,st) = \delta(u,s)\ \delta(u\cdot s,t)$ for $u \in r(st)$

$\delta(u,s^{-1}) = |\delta(u\cdot s^{-1},s)|^{-1}$ for $u \in d(s)$

<u>3.27. Definition :</u> Let G be a locally compact groupoid with Haar system. We will say that G has <u>sufficiently many non-singular Borel G-sets</u> if for every measure μ on G^0 with induced measure ν on G, every Borel set in G of positive ν-measure contains a non-singular Borel G-set s of positive μ-measure, that is, such that $\mu(r(s)) > 0$.

<u>3.28. Examples :</u>

(a) Transformation group. Let μ be a measure on the unit space U of the transformation group (U,S). A Borel subset of $U \times S$ of positive $\mu \times \lambda$-measure, where λ is a left Haar measure for S, contains a rectangle $A \times B$ with A,B Borel, $\mu(A) > 0$ and $\lambda(B) > 0$. Choose $s \in B$. Then $s = \{(u,s) : u \in A\}$ is a non-singular Borel G-set of positive μ-measure.

(b) r-discrete groupoids. Let μ be a measure on the unit space of a second countable r-discrete groupoid G. Let E be a Borel set in G of positive ν-measure. Since G can be covered by countably many open G-sets, there exists an open G-set t such that $s = E \quad t$ has positive ν-measure. Then, s is a non-singular Borel G-set of positive μ-measure.

(c) Transitive principal groupoids. Let X be a locally compact space. We define the transitive groupoid on the space X as $G = X \times X$, with the groupoid structure given in 1.2.c and the product topology. We know that a Haar system on G is defined by a measure α of support X. If X is uncountable and satisfies the second axiom of countability, and if α is non-atomic, then G has sufficiently many non-singular Borel G-sets. This can be seen as follows : there is a Borel isomorphism of X onto $\mathbb{R}$ carrying α into the Lebesgue measure. Thus the problem is reduced to the case $X = \mathbb{R}$, α = Lebesgue measure. Then the transitive groupoid is isomorphic to the groupoid of the transformation group $(\mathbb{R}, \mathbb{R})$ where $\mathbb{R}$ acts by translation and we may conclude by a.

Question : Assume that G has sufficiently many non-singular Borel G-sets and that μ is a measure on G^0 quasi-invariant under every non-singular Borel G-sets ; can we conclude that μ is quasi-invariant ?

The existence of sufficiently many non-singular Borel G-sets will be needed in the second chapter (theorem 2.1.21).

4. Continuous Cocycles and Skew-Products

The asymptotic range of a continuous one-cocycle (definition 4.3) is used to solve a few problems concerning the triviality of cocycles and the irreducibility of skew-products. This section closely follows [56], [57] and [58] where a similar study has been done for C^*-algebras.

Let G be a topological groupoid (definition 2.1). If E is a subset of the unit space G^0, [E] will denote its saturation : $[E] = r[d^{-1}(E)]$. If $E = [E]$, we say that E is invariant (or invariant under G if there is any ambiguity). We will always assume that the range map $r : G \to G^0$ is open. Recall (2.4) that locally compact groupoids with a left Haar system have this property. Then, the saturation of an open subset of G^0 is open.

4.1. Definition : Let G be a topological groupoid with open range map.

(i) G is minimal if the only open invariant subsets of G^0 are the empty set $\emptyset$ and G^0 itself.

(ii) G is irreducible if every non-empty invariant open subset of G^0 is dense.

If there exists a dense orbit, then G is irreducible. The converse holds if G is second countable and locally compact. It is useful to note that the irreducibility of G may be expressed as the density of the image of G in $G^0 \times G^0$ by the map (r,d) : $G \to G^0 \times G^0$, $x \to (r(x),d(x))$. These notions of minimality and irreducibility could have been defined in terms of the structure space $G^0//G$ of G, obtained from the quotient space G^0/G by identifying orbits with the same closure, but we will not make use of it here. The next proposition shows that they are invariant under continuous similarity.

4.2. Proposition : Suppose that G and H are topological groupoids which are continuously similar, that is, which are similar as in definition 1.3 where the homomorphisms $\phi : G \to H$ and $\Psi : H \to G$ are continuous. Then the map $0 \to (\phi^0)^{-1}(0)$ sets up a bijection between the invariant open subsets of H and G.

Proof : Let 0 be an invariant open subset of H. The $(\phi^0)^{-1}(0)$ is open and invariant. For, if $x \in G$ and $\phi^0[d(x)] \in 0$, then $\phi^0[r(x)] = r[\phi^0(x)] \in 0$ since 0 is invariant.

Moreover, $(\psi^0 \circ \phi^0)^{-1}(0) = 0$.

Indeed $(\psi \circ \phi)(x) = (\theta \circ r)(x)\ x\ (\theta \circ d(x))^{-1}$

$\psi^0 \circ \phi^0(u) = r[\theta(u)]$ with $d[\theta(u)] = u$

therefore $u \in 0$ iff $\psi^0 \circ \phi^0(u) \in 0$.

Q.E.D.

Let G be a topological groupoid and A a topological group, not necessarily abelian. We may still define (cf 1.11) the following objects. The set of continuous homomorphisms from G to A is denoted by $Z^1(G,A)$. The subset of $Z^1(G,A)$ consisting of elements of the form $c(x) = [b\circ r(x)][b\circ d(x)]^{-1}$ where b is a continuous function from G^0 to A is denoted by $B^1(G,A)$. Moreover, we say that two elements c and c' in $Z^1(G,A)$ are cohomologous if there exists a continuous function b from G^0 to A such that $c'(x) = [b\circ r(x)]\ c(x)\ [b\circ d(x)]^{-1}$.

The following definition of the asymptotic range of a cocycle is the topological version of the definition 8.2 of [31,I].

4.3. Definition : Let G be a topological groupoid, A a topological group and c an element of $Z^1(G,A)$.

(i) The range of c is $R(c)$ = closure of $c(G)$.

(ii) The asymptotic range of c is $R_\infty(c) = \cap R(c_U)$, where the intersection is taken over all non-empty open subsets U of G^0 and c_U denotes the restriction of c to $G_{|U}$. Moreover, let u be a unit of G.

(iii) The range of c at u is $R^u(c)$ = closure of $c(G^u)$.

(iv) The asymptotic range of c at u is $R^u_\infty = \cap R^u(c_U)$, where the intersection is taken over a base of neighborhoods of u.

We use in the following definition the character group $\hat{A}$ of a topological group A ; it is the group of continuous homomorphisms of A into the circle group $\mathbb{T}$.

4.4. Definition : Let G be a topological groupoid, A a topological group and c an element of $Z^1(G,A)$. The T-set of c is $T(c) = \{\chi \in \hat{A} : \chi \circ c \in B^1(G,\mathbb{T})\}$.

The following proposition gives some basic properties of the quantities $R_\infty(c)$ and $T(c)$; in particular, they depend only on the cohomology class of c. The aim of this section is to show their usefulness, justifying their introduction. Further references to the asymptotic range and the T-set of a cocycle can be found in [31] in the context of ergodic theory. It is interesting to note that they were first introduced on a work about operator algebras, namely, the Araki-Woods classification of factors obtained as infinite tensor products of factors of type I.

4.5. Proposition : Let G be a topological groupoid with open range map, A a topological group and $c \in Z^1(G,A)$. Then

(i) $R_\infty(c)$ is a closed subgroup of A, $T(c)$ is a subgroup of $\hat{A}$, and $R_\infty(c)$ and $T(c)$ are orthogonal to each other.

(ii) $R_\infty(c)$ and $T(c)$ depend only on the class of c.

(iii) $R_\infty(e) = \{e\}$ and $T(e) = \hat{A}$, where e denotes the identity element of A as well as the constant cocycle $e(x) = e$.

Proof :

(i) Let us first show that $R(c)\, R_\infty(c) \subset R(c)$. Suppose $a \in R(c)$ and $b \in R_\infty(c)$. For every neighborhood V of b, $r[c^{-1}(V)]$ is dense in G^0 : if not, there would exist a non-empty open subset O avoiding $r[c^{-1}(V)]$ and $c_0^{-1}(V)$ would be empty. Let W be a neighborhood of ab and choose U,V open neighborhoods of a and b respectively such that $UV \subset W$. Since $d[c^{-1}(U)]$ is a non-empty open set and $r[c^{-1}(V)]$ is dense, there exist $x, y \in G$ such that $c(x) \in U$, $c(y) \in V$ and $d(x) = r(y)$. Then, $c(xy) = c(x)c(y) \in UV \subset W$. This shows $ab \in R(c)$. We deduce that $R_\infty(c)$ is stable under multiplication : for any non-empty open set U of G^0, $R_\infty(c)\, R_\infty(c) \subset R(c_U)\, R_\infty(c_U) \subset R(c_U)$ hence $R_\infty(c)\, R_\infty(c) \subset R_\infty(c)$. As it is closed, symmetric and contains e, $R_\infty(c)$ is a closed subgroup of A.

Since $B^1(G,\mathbb{T})$, with pointwise multiplication, is a group, $T(c)$ is a subgroup of $\hat{A}$. We finally have to show that for every $\chi \in T(c)$ and every $a \in R_\infty(c)$, $\chi(a) = 1$. For every closed neighborhood V of 1 in $\mathbb{T}$, there exists a non-empty open set U in G^0 such that $(\chi \circ c)(G_U) \subset V$ because $\chi \circ c \in B^1(G,\mathbb{T})$; in particular, $\chi(a) \in V$.

(ii) Suppose that $c'(x) = [b \circ r(x)] c(x) [b \circ d(x)]^{-1}$ with $c \in Z^1(G,A)$ and b a continuous map from G^0 to A. Let $a \in R_\infty(c)$. We want to show that $a \in R_\infty(c')$; that is, given a non-empty open set U' on G^0 and a neighborhood W' of a, we want to show that $W' \cap c'(G_{|U'}) \neq \emptyset$. We choose $u \in U'$, a neighborhood V of $b(u)$ and a neighborhood W of a such that $VWV^{-1} \subset W'$. There exists an open neighborhood U of u such that $b(U) \subset V$. Since $W \cap c(G_{|U}) \neq \emptyset$, we are done. We have shown $R_\infty(c) \subset R_\infty(c')$, hence $R_\infty(c) = R_\infty(c')$. The equality $T(c) = T(c')$ results from the definition of a T-set.

(iii) Clear.

Q.E.D.

Similar proofs yield similar results about the asymptotic range of a cocycle at a unit u.

<u>4.6. Proposition :</u> Let G, A, c be as before and $u \in G^0$. Then

(i) $R^u(c)\ R^u_\infty(c) = R^u(c)$.

(ii) $R^u_\infty(c)$ is a closed subsemi-group of A

(iii) $R^u_\infty(c)$ depends only on the class of c.

(iv) $R^u_\infty(e) = \{e\}$

(v) If $u \sim v$, $R^u(c) = R^v(c)$.

To proceed further, an additional assumption on the topological groupoid G will be needed. Let us recall the definition 3.24.i ;

<u>4.7. Definition :</u> Let G be a topological groupoid. A G-set s (definition 1.10) will be called a <u>continuous G-set</u> if the restriction of r and d to s is a homeomorphism onto an open subset of G^0.

An open G-set of an r-discrete locally compact groupoid with Haar measure is a continuous G-set. For another example, consider the groupoid of a topological transformation group (U,S) ; let V be an open subset of U and $s \in S$; then the G-set $s = \{(u,s) : u \in V\}$ is a continuous G-set. In both examples, the groupoid admits a cover of continuous G-sets. This is the assumption we need.

4.8. Proposition : Let G be a topological groupoid, A a topological abelian group and $c \in Z^1(G,A)$.

(i) If $c \in B^1(G,A)$, then for any neighborhood V of e in A and any $u \in G^0$, there exists an open neighborhood U of u such that $R(c_U) \subset V$.

(ii) If G admits a cover of continuous G-sets, if G^0 is compact and if there exists a dense orbit, then the converse holds.

Proof :

(i) Clear since $c(x) = b\circ r(x) - b\circ d(x)$.

(ii) We assume that c satisfies the condition that for any neighborhood V of e in A and any $u \in G^0$, there exists an open neighborhood of u such that $c(G_{|U}) \subset V$. This means in particular that c vanishes on the isotropy group bundle of G. Let us introduce the principal groupoid associated with G. It is determined by the equivalence relation $\sim$ on G^0. As a set, it is the image of the map $(r,d) : G \to G^0 \times G^0$. We provide it with the final topology, which is usually strictly finer than the topology induced from $G^0 \times G^0$. The cocycle c factors through the map (r,d) : $c(x) = c'(r(x),d(x))$. Let V be a neighborhood of 0 in A. Since $\{U$ non-empty open set in G^0 such that $c'(U\times U) \quad V\}$ is an open cover of G^0, there exists a finite subcover hence an entourage $\mathcal{U}$ of the uniformity on G^0 such that

$$(u,v) \in \mathcal{U} \text{ and } u \sim v \Longrightarrow c'(u,v) \in V.$$

Let us show that c' is continuous with respect to the topology induced from $G^0 \times G^0$. Given $(u,v) \in G^0 \times G^0$ with $u \sim v$, let $x \in G$ be such that $r(x) = u$ and $d(x) = v$ and let s be a continuous G-set containing x. Consider (u',v') with $u' \sim v'$. For u' sufficiently close to u, there exists $y \in s$ with $r(y) = u'$ and $c'(u',w) - c'(u,v)$, where $w = d(y)$, can be made arbitrarily small. On the other hand $c'(u',v') - c'(u',w)$ $= c'(w,v')$ can be made arbitrarily small, provided that v' is close enough to w, this happens if u' is sufficiently close to u and v' sufficiently close to v.

Next we show that c' is uniformly continuous on the dense subset $[u_0] \times [u_0]$ of $G^0 \times G^0$, where u_0 has a dense orbit. If u, v, u', v' are in the orbit of u_0, then $c'(u',v') - c'(u,v) = c'(u',u) - c'(v',v)$. Therefore, c' extends to a continuous function on $G^0 \times G^0$. Then, $f(u) = c'(u,u_0)$ is a continuous function on G^0 and c' agrees with its coboundary on $[u_0] \times [u_0]$, hence on G.

<u>4.9. Proposition :</u> Let G be a topological groupoid admitting a cover of continuous G-sets, A a topological abelian group and $c \in Z^1(G,A)$. Suppose $u_0 \in G^0$ has a dense orbit. Then $R_\infty(c) = R(c_U)$, where the intersection is taken over a base of neighborhoods of u_0.

<u>Proof :</u> Suppose that $a \in R(c_U)$ for every U in a base of neighborhoods of u_0. Let V,W be neighborhood of e on A such that $W + W \subset V$ and U be a non-empty open set. There exists $x \in G$ with $r(x) = u_0$, $d(x) \in U$ and a continuous G-set s containing x. We may assume that $d(s) \subset U$ and $c(s) - c(s) \subset W$. Because $a \in R(c_{r(s)})$, there exists $y \in G_{|r(s)}$ such that $c(y) \in a + W$. Let $z = s^{-1}ys$, then $z \in G_{|d(s)} \subset G_{|U}$ and

$$c(z) = c(s^{-1}r(y)) + c(y) + c(d(y)s) \in a + W + W \subset a + V.$$

Thus, $a \in R(c_U)$ for any non-empty open set U.

Q.E.D.

The following theorem may be compared with theorem 9 of [31,I]. Combined with the results of the second chapter, it yields a particular case of a well-known theorem of Sakai which states that every bounded derivative of a simple C^*-algebra with identity is inner.

<u>4.10. Theorem :</u> Let G be a topological groupoid admitting a cover of continuous G-sets and a compact unit space, let A be a topological abelian group and let $c \in Z^1(G,A)$. Assume that G is minimal. If R(c) is compact and $R_\infty(c) = \{0\}$, then $c \in B^1(G,A)$.

<u>Proof :</u> We use 4.8 (ii). Suppose that there exists an open neighborhood V of 0 in A, $u \in G^0$, a base of neighborhoods of u and a net $\{x_U\}$ such that

$$x_U \in G_{|U} \text{ and } c(x_U) \notin V.$$

If $\{a_U\}$ is a subset of $\{c(x_U)\}$ converging to a, then $a \notin V$ and $a \in \cap R(c_U)$ where the intersection is taken over a base of neighborhoods of u. By 4.9, $a \in R_\infty(c)$. Since $R_\infty(c) = \{0\}$, this is a contradiction.

Q.E.D.

Note that if A is torsion free, the condition R(c) compact already implies $R_\infty(c) = \{0\}$.

The next theorem may be compared with théorème 2.3.1 of [13] in the context of von Neumann algebras and with theorem 4.2 of [56] in the context of C^*-algebras. The proof is adapted from [56].

<u>4.11. Theorem :</u> Let G be a topological groupoid admitting a cover of continuous G-sets and a compact unit space, let A be a locally compact abelian group and let $c \in Z^1(G,A)$. Assume that G is minimal, then if $R(c)/R_\infty(c)$ is compact in $A/R_\infty(c)$, it follows that T(c) is the annihilator of $R_\infty(c)$ in $\hat{A}$.

<u>Lemma</u> : Let G be a topological groupoid admitting a cover of continuous G-sets, let A be a locally compact abelian group and let $c \in Z^1(G,A)$. Assume that G is irreducible. Then $\mathcal{B} = \{V + R(c_U) :$ V compact neighborhood of 0 in A and U non-empty open subsets of $G^0\}$ is a base of a filter. Its intersection is $R_\infty(c)$.

<u>Proof :</u> As in 3.4. of [56], it suffices to show that given a compact neighborhood V of 0 in A and non-empty open subsets U_i of G^0, i = 1,2, there exist non-empty open subsets $U_i' \subset U_i$, i = 1,2 such that $R(c_{U_i'}) \subset V + R(c_{U_j'})$ i,j = 1,2 and $i \neq j$. We choose $x \in G$ with $r(x) \in U_1$ and $d(x) \in U_2$ and a continuous G-set s containing x. We may assume that $r(s) \subset U_1$, $d(s) \subset U_2$ and $c(s) - c(s) \subset V$. Then $U_1' = r(s)$ and $U_2' = d(s)$ will do.

<u>Proof of the theorem :</u> With the notations of the lemma, the image of $\mathcal{B}$ in $A/R_\infty(c)$ is a base of a filter of compact sets with intersection {0}. Hence, given a neighborhood V of 0 in A, we may find a non-empty open set U in G^0 such that $R(c_U) \subset V + R_\infty(c)$. Thus, if χ is orthogonal to $R_\infty(c)$, $R_\infty(\chi \circ c) = \{1\}$. By 4.10, $\chi \circ c \in B^1(G,\mathbb{T})$, that is, $\chi \in T(c)$. The reverse inclusion has been shown in 4.5. (i).

Recall that, given a groupoid G, a group A and $c \in Z^1(G,A)$, one may define the skew-product G(c), whose underlying space is G x A and unit space is $G^0 \times A$. If G and A are topological and c continuous, G(c) with the topology of G x A is a topological groupoid. Note that if G has an open range map [resp. a cover of continuous G-sets], then so has G(c).

The following characterization of the asymptotic range of a cocycle in terms of the skew-product is taken from Pedersen [60] 8.11.8. It will be used in Section 5 of Chapter 2. Recall that there is a canonical action of A on the skew-product G(c), given by

$$(x,b) \cdot a = (x,a^{-1}b)$$

<u>4.12. Proposition :</u> Let G be a topological groupoid with open range map, let A be a topological abelian group and let $c \in Z^1(G,A)$. Then the following properties are equivalent for $a \in A$:

(i) $a \in R_\infty(c)$ and

(ii) for any non-empty open invariant subset 0 of the unit space G(c), $0 \cap 0 \cdot a$ is non-empty.

<u>Proof :</u>

(i) $\Longrightarrow$ (ii) Suppose $a \in R_\infty(c)$.

Let 0 be a non-empty invariant subset of $G^0 \times A$. It contains a non-empty rectangle $U \times V$, with U open on G^0 and V open in A. Let $b \in V$. Since $a \in R_\infty(c)$, there exists $x \in G_{|U}$ such that $c(x) \in a - b + V$. Then $(r(x),b)$ and $(d(x), b - a + c(x))$ belong to $U \times V \cap 0$. Since $(r(x), b - a)$ is equivalent to $(d(x), b - a + c(x))$, it belongs to 0. Since $(r(x), b - a) = (r(x), b) \cdot a$, it also belongs to $0 \cdot a$.

(ii) $\Longrightarrow$ (i) Suppose that a satisfies (ii).

Let U be a non-empty open set in G^0 and V be a neighborhood of 0 in A. Choose a neighborhood W of 0 such that $W - W \subset V$. Since the saturation of $U \times W$ in the unit space of G(c) is an invariant open set, it contains an element (v,b) together with (v,b-a). This implies the existence of x and y in G such that

$r(x) = v$ and $(d(x),b + c(x)) \in U \times W$ and

$r(y) = v$ and $(d(y), b - a + c(y)) \in U \times W$.

Then, $x^{-1}y \in G_{|U}$ and $c(x^{-1}y) = -c(x) + c(y) \in a + W - W \subset a + V$.

This shows that $a \in R_\infty(c)$.

Q.E.D.

4.13. Proposition : Let G be a topological groupoid with open range map, let A be a topological group and let $c \in Z^1(G,A)$. The following properties are equivalent :

(i) G is irreducible and $R_\infty(c) = A$ and

(ii) G(c) is irreducible.

Proof :

(i) $\Longrightarrow$ (ii) It suffices to show that, given non-empty open sets U_1, U_2 in G^0, a neighborhood V of e in A and $a \in A$, there exists $z \in G$ such that $r(z) \in U_1$, $d(z) \in U_2$ and $c(z) \in aV$. Choose W, open neighborhood of e such that $W^{-1}W \subset V$. Since G is irreducible, there exists $b \in A$ such that $c^{-1}(bW) \cap r^{-1}(U_1) \cap d^{-1}(U_2) \neq \emptyset$. Let $U = r[c^{-1}(bW) \cap r^{-1}(U_1) \cap d^{-1}(U_2)]$. Since $ba^{-1} \in R_\infty(c)$, there exists $x \in G_{|U}$ such that $c(x) \in bWa^{-1}$. Since $r(x) \in U$, there exists $y \in G$ such that $r(y) = r(x)$, $d(y) \in U_2$ and $c(y) \in bW$. Let $z = x^{-1}y$. Then $r(z) = d(x) \in U \subset U_1$, $d(z) = d(y) \in U_2$ and $c(z) = c(x)^{-1}c(y) \in a W^{-1}W \subset aV$.

(ii) $\Longrightarrow$ (i) If G(c) is irreducible, then G is clearly irreducible. To show that $R_\infty(x) = A$, let $a \in A$, let V and W be neighborhoods of e in A such that $W^{-1}W \subset V$ and let U be a non-empty open subset of G^0. Since G(c) is irreducible, there exists $x \in G_U$ and $b \in W$ such that $bc(x) \in Wa$. Then $c(x) \in W^{-1}Wa \subset Va$. This shows that $a \in R_\infty(c)$.

Q.E.D.

4.14. Proposition : Let G be a topological groupoid with open range map, let A be a topological group and let $c \in Z^1(G,A)$. Let $(u,a) \in G^0 \times A$.

(i) If (u,a) has a dense orbit relative to G(c), then the asymptotic range of c at u, $R_\infty^u(c)$, is equal to A.

(ii) Conversely, if G is minimal and if $R_\infty^u(c) = A$, then (u,a) has a dense orbit.

Proof :

(i) Suppose that the orbit $[(u,a)] = \{(d(x),ac(x)) : x \in G^u\}$ is dense in $G^0 \times A$. Let $b \in A$, let V be a neighborhood of b and let U be an open neighborhood of u. There exists $x \in G^u$ such that $(d(x), ac(x)) \in U \times aV$. Thus, $x \in G^u \cap G_{|U}$ and $c(x) \in V$. We conclude that $b \in R^u_\infty(c)$.

(ii) Suppose that $R^u_\infty(c) = A$. Let F be the closure of the orbit of (u,a). For any $b \in A$, $(u,b) \in F$: indeed, let U be an open neighborhood of u and V a neighborhood of b ; since $a^{-1}b \in R^u_\infty(c)$, there exists x such that $r(x) = u$, $d(x) \in U$ and $c(x) \in a^{-1}V$; in other words, $(d(x),ac(x)) \in U \times V$. The set $\{v \in G^0 :$ for any $b \in A$, $(v,b) \in F\}$ is non-empty, G-invariant and closed. Since G is minimal, this is G^0, hence $F = G^0 \times A$.

Q.E.D.

4.15. Proposition : Let G be a topological groupoid with open range map, A a topological group and $c \in Z^1(G,A)$. Assume that A is compact, then $R_\infty(c) = R^u_\infty(c)$ for every $u \in G^0$ with a dense orbit.

Proof : We first show that $R(c) = R^u(c)^{-1}R^u(c)$ for u with a dense orbit. The inclusion $R^u(c)^{-1} R^u(c) \subset R(c)$ holds for arbitrary u. Suppose now that $a \in R(c)$ and u has a dense orbit. Since A is compact, it suffices to show that a belongs to the closure of $R^u(c)^{-1} R^u(c)$. If V is a neighborhood of a, $r[c^{-1}(V)] \cap [u]$ is non-empty : there exist x,y such that $c(x) \in V$, $r(x) = d(y)$ and $r(y) = u$. Then, $c(y)^{-1} c(yx) \in [c(G^u)^{-1} c(G^u)] \cap V$. Therefore $R(c_U) = R^u(c_U)^{-1} R^u(c_U)$ for any open neighborhood U of u. Using the compactness of A, one may write :

$$R_\infty(c) = \cap R(c_U) = [\cap R^u(c_U)]^{-1} [\cap R^u(c_U)] = R^u_\infty(c)^{-1} R^u_\infty(c) = R^u_\infty(c)$$

where the intersections are taken over all open neighborhoods of u. The last equality holds because, in a compact group, any closed semi-group is a group.

Q.E.D.

4.16. Corollary : Let G be a topological groupoid with open range map, let A be a topological group and let $c \in Z^1(G,A)$.

(i) If $G(c)$ is minimal, then G is minimal and $R_\infty(c) = A$

(ii) If A is compact, if G is minimal, and if $R_\infty(c) = A$, then $G(c)$ is minimal.

Proof :

(i) If $G(c)$ is minimal, G is clearly minimal. Moreover, $R_\infty(c) = A$ by 4.12.

(ii) Using 4.14 and 4.13 (ii), one sees that every $(u,a) \in G^0 \times A$ has a dense orbit.

Q.E.D.

4.17. Proposition : Let G be a topological groupoid with open map, let A be a group with the discrete topology and let $c \in Z^1(G,A)$. The following properties are equivalent :

(i) G is irreducible and $R_\infty(c) = R(c)$; and

(ii) $c^{-1}(e)$ is irreducible.

Proof :

(i) $\Rightarrow$ (ii) Let U_1 and U_2 be non-empty open sets in G^0. By irreducibility of G, there is $a \in A$ such that $c^{-1}(a) \cap r^{-1}(U_1) \cap d^{-1}(U_2)$ is non-empty. Then $U = r[c^{-1}(a) \cap r^{-1}(U_1) \cap d^{-1}(U_2)]$ is a non-empty open set and since $a^{-1} \in R_\infty(c)$, there exists $x \in G_U$ with $c(x) = a^{-1}$. Therefore, there is $y \in G$ such that $d(x) = r(y)$, $c(y) = a$ and $d(y) \in U_2$. Consider $z = xy$: $d(z) \in U_2$, $r(z) = r(x) \in U_1$ and $c(z) = c(x)c(y) = e$. This shows that the groupoid $c^{-1}(e)$ is irreducible.

(ii) $\Rightarrow$ (i) If $c^{-1}(e)$ is irreducible, so is G. Consider $a \in R(c)$ and U a non-empty open subset of G^0. Since $c^{-1}(e)$ is irreducible, $c^{-1}(e) \cap r^{-1}(U) \cap d^{-1}[r(c^{-1}(a))] = V$ is a non-empty open set and so is $c^{-1}(e) \cap r^{-1}[d(V)] \cap d^{-1}(U)$. Therefore, we can find x, y, z such that : $c(x) = e$, $c(y) = a$, $c(z) = e$, $d(x) = r(y)$, $r(z) = d(y)$, $r(x) \in U$ and $d(z) \in U$. Then, $xyz \in G_{|U}$ and $c(xyz) = a$. This shows that $a \in R_\infty(c)$.

Q.E.D.

Another subgroup of A can be attached to a cocycle $c \in Z^1(G,A)$ (cf. [62], theorem of Section 2). We conclude this section by discussing briefly how it is related to $R_\infty(c)$ and $T(c)^\perp$ (the annihilator of $T(c)$ in A) in a particular case.

4.18. Definition : Let G be a topological groupoid, let A be a topological group and let $c \in Z^1(G,A)$. We define $R_1(c)$ to be the set of elements a of A with the property that for every $G(c)$-invariant complex-valued continuous function on $G^0 \times A$

and for every (u,b) in $G^0 \times A$, the equality $f(u,ba) = f(u,b)$ holds.

4.19. Proposition : Let G be a topological groupoid, let A be a topological group and let $c \in Z^1(G,A)$. Then $R_\infty(c) \subset R_1(c) \subset T(c)^\perp$.

Proof : $R_\infty(c) \subset R_1(c)$. If $a \notin R_1(c)$, there is a continuous function f on $G^0 \times A$, which is $G(c)$ invariant, and $(u,b) \in G^0 \times A$ such that $f(u,ba) \neq f(u,b)$, hence there exist an open neighborhood U of u and a neighborhood V of a such that $f(U \times bV) \cap f(U \times bVa^{-1}) = \emptyset$. If $x \in G_U$ and $c(x) \in V$, then $f(r(x),b) = f(d(x),bc(x))$. This is a contradiction and therefore $a \notin R_\infty(c)$.

$R_1(c) \subset T(c)^\perp$. Let $a \in R_1(c)$ and $\chi \in T(c)$, that is $\chi \circ c \in B^1(G,\mathbb{T})$. Then, there exists $g : G^0 \to \mathbb{T}$ continuous such that $g \circ d(x)\ \chi \circ c(x) = g \circ r(x)$ for every $x \in G$. Let $f(u,b) = g(u)\ \chi(b)$. Then f is continuous and $G(c)$-invariant. Therefore, $f(u,ba) = f(u,b)$ that is, $g(u)\chi(b)\chi(a) = g(u)\chi(b)$, hence $\chi(a) = 1$.

Q.E.D.

More information can be obtained in the case of a compact abelian group A.

4.20. Proposition : Let G be a topological groupoid, let A be a topological group and let $c \in Z^1(G,A)$. Assume that G is minimal and A is compact and abelian. Then $R_1(c) = T(c)^\perp$.

Proof : Let f be a continuous $G(c)$-invariant function on $G^0 \times A$. For each $\chi \in \hat{A}$,

$g(u) = \int f(u,a)\ \chi(a)da$ is continuous and satisfies

$g \circ d(x)\ \ \chi \circ c(x) = g \circ r(x)$

Since G is minimal, either g vanishes identically or not at all and, in the latter case, $\chi \circ c \in B^1(G,\mathbb{T})$, that is, $\chi \in T(c)$. Thus, for every u, the Fourier transform of $f(u,\cdot)$ is supported on $T(c)$. Hence, if $a \in T(c)^\perp$, then for any $b \in A$

$f(u, a + b) = f(u,b)$. So $a \in R_1(c)$.

Q.E.D.

We also recall that under the hypotheses of 4.11, $R_\infty(c) = R_1(c) = T(c)^\perp$. These last facts, combined with 4.15, give a theorem of Rauzy ([62], theorem of section 2) about the minimality of a skew-product.

CHAPTER II

THE C^*-ALGEBRA OF A GROUPOID

First, let us say that "groupoid" stands for locally compact groupoid with a fixed Haar system (definition 1.2.2) chosen once for all. We shall see (corollary 2.11) how the C^*-algebra can be affected by another choice of Haar system. We also assume that the topology of the groupoid is second countable.

The goal here is to construct the C^*-algebra of a groupoid in a way which extends the well-known cases of a group (e.g. Dixmier [19]) or of a transformation group (e.g. Effros-Hahn [23]). In fact, our construction closely follows [23]: the space $C_c(G)$ of continuous functions with compact support is made into a $*$-algebra and endowed with the smallest C^*-norm making its representations continuous ; $C^*(G)$ is its completion. The details are in Section 1. We refrain from putting any modular function in the definition of the involution, since none is available. However, this is a minor change and the C^*-algebra so obtained is isomorphic to the usual one in the case of a transformation group. Let us note that, in the case of a transformation group, the $*$-algebra $C_c(G)$ has been studied by Dixmier ([16],§ X) in the context of quasi-unitary algebras.

If σ is a continuous 2-cocycle on G with values in the circle group, the C^*-algebra $C^*(G,\sigma)$ is defined in the same fashion. One of the main justifications for its introduction, besides the need to deal with projective representations, is given in Section 4, where the C^*-algebra of an r-discrete principal groupoid is characterized, under suitable conditions, by the existence of a particularly nice kind of maximal abelian subalgebra. One of these conditions is amenability, which is defined in Section 3.

An essential tool in the study of the C^*-algebra of a groupoid is the correspondence, very familiar in the case of a group, between the unitary representations of the groupoid and the non-degenerate representations of the C^*-algebra. It is established at the end of Section 1 and under a condition (existence of sufficiently many non-singular G-sets) sufficient for our applications.

Of particular interest are the regular representations of a groupoid. They have been studied extensively since Murray and Von Neumann and we refer to Hahn [45] for further details. They appear under various forms, one of them is as representations induced from the unit space ; this is described in Section 2, where the inducing process from more general subgroupoids is also considered.

The last section, Section 5, interprets in the language of C^*-algebras the results of Section 4 of the first chapter. They center around the question of primitivity and simplicity of a crossed-product algebra.

1. The Convolution Algebras $C_c(G,\sigma)$ and $C^*(G,\sigma)$

Let G be a locally compact groupoid with left Haar system $\{\lambda^u\}$ and let σ be a continuous 2-cocycle in $Z^2(G,\mathbb{T})$. For f and g $\in C_c(G)$, let us define

$$f*g(x) = \int f(xy)g(y^{-1})\,\sigma(xy,y^{-1})d\lambda^{d(x)}(y) \;,$$

$$f^*(x) = \overline{f(x^{-1})}\;\sigma(x,x^{-1}).$$

1.1. Proposition : Under these operations, $C_c(G)$ becomes a topological $*$-algebra, denoted by $C_c(G,\sigma)$.

Proof : We first show that these operations are well defined. For each x, $f*g(x)$ is the value of the integral of a continuous function with compact support. Since $f*g(x)$ is nonzero only if there is y such that f(xy) and $g(y^{-1})$ are nonzero, supp$(f*g)$ is contained in the compact set (suppf)(suppg). To show the continuity of $f*g$, we may use the same device as Connes in [14] 2.2. That is, since G^2 is a closed subset of the normal space G x G, the function $(x,y) \to f(xy)\sigma(xy,y^{-1})$ may be extended to a bounded continuous function k on G x G. Since the function

$$x \mapsto \ell_x : G \to C_c(G),$$

where $\ell_x(y) = k(x,y)g(y^{-1})$, is continuous, so is the function $(x,u) \to \int k(x,y)g(y^{-1})d\lambda^u(y) : G \times G^0 \to \mathbb{C}$; in particular, its restriction to $(x,d(x))$ is continuous. Note that f^* is also continuous, with compact support $\mathrm{supp} f^* = (\mathrm{supp} f)^{-1}$. The convolution is associative : if $f, g, h \in C_c(G)$,

$$\begin{aligned} f*(g*h)(x) &= \int f(xy)\, g\, h(y^{-1})\, \sigma(xy,y^{-1})d\lambda^{d(x)}(y) \\ &= \iint f(xy)\, g(y^{-1}z)\, h(z^{-1})\, \sigma(y^{-1}z,z^{-1})\sigma(xy,y^{-1})d\lambda^{r(y)}(z)d\lambda^{d(x)}(y) \\ &= \iint f(xy)\, g(y^{-1}z)\, h(z^{-1})\, \sigma(xy,y^{-1})\, \sigma(y^{-1}z,z^{-1})d\lambda^{r(z)}(y)d\lambda^{d(x)}(z) \\ &= \iint f(xzy)\, g(y^{-1})\, h(z^{-1})\, \sigma(xzy,y^{-1}z^{-1})\sigma(y^{-1},z^{-1})d\lambda^{d(z)}(y)d\lambda^{d(x)}(z) \\ &= \iint f(xzy)\, g(y^{-1})\, h(z^{-1})\, \sigma(xzy,y^{-1})\, \sigma(xz,z^{-1})d\lambda^{d(z)}(y)d\lambda^{d(x)}(z) \\ &= \int (f*g)(xz)\, h(z^{-1})\, \sigma(xz,z^{-1})d\lambda^{d(x)}(z) \\ &= (f*g)*h\,(x). \end{aligned}$$

The involution is involutive :

$$f^{**}(x) = \overline{f^*(x^{-1})}\ \overline{\sigma(x,x^{-1})} = f(x)\, \sigma(x^{-1},x)\, \overline{\sigma(x,x^{-1})} = f(x).$$

Also

$$(f*g)^*(x) = \overline{f*g(x^{-1})}\ \overline{\sigma(x,x^{-1})} = \int \overline{f(x^{-1}y)}\ \overline{g(y^{-1})}\ \overline{\sigma(x^{-1}y,y^{-1})}\ \overline{\sigma(x,x^{-1})}d\lambda^{r(x)}(y).$$

Using $\overline{\sigma(x^{-1}y,y^{-1})} = \overline{\sigma(y,y^{-1})}\ \sigma(x^{-1},y)$

and $\overline{\sigma(x,x^{-1})} = \overline{\sigma(x^{-1},y)}\ \overline{\sigma(x^{-1}y,y^{-1}x)}\ \overline{\sigma(y,y^{-1}x)}$, we obtain

$$\begin{aligned} (f*g)^*(x) &= \int \overline{g(y^{-1})}\ \overline{\sigma(y,y^{-1})}\ \overline{f(x^{-1}y)}\ \overline{\sigma(x^{-1}y,y^{-1}x)}\ \sigma(y,y^{-1}x)\, d\lambda^{r(x)}(y) \\ &= \int g^*(y)\, f^*(y^{-1}x)\, \sigma(y,y^{-1}x)d\lambda^{r(x)}(y) \\ &= g^* * f^*(x). \end{aligned}$$

Finally, the operations are continuous. If $f_n \to f$ and $g_m \to g$, there exist compact sets K and L such that, eventually, $\mathrm{supp}\, f_n \subset K$ and $\mathrm{supp}\, g_m \subset L$. Then, $\mathrm{supp}\, f_n * g_m \subset KL$. Also,

$$\begin{aligned} |f*g(x) - f_n * g_m(x)| &\le \int |f(xy)g(y^{-1}) - f_n(xy)g_m(y^{-1})|d\lambda^{d(x)}(y) \\ &\le \int |f(xy) - f_n(xy)|\,|g(y^{-1})|d\lambda^{d(x)}(y) \\ &\quad + \int |f_n(xy)|\,|g(y^{-1}) - g_m(y^{-1})|\,d\lambda^{d(x)}(y) \end{aligned}$$

Therefore, $f_n * g_m$ converge uniformly to $f * g$ on KL. Moreover, supp $f_n \subset K^{-1}$ and $|f_n^*(x) - f^*(x)| = |f_n(x^{-1}) - f(x^{-1})|$ converges uniformy to zero on K^{-1}.

Q.E.D.

1.2. Proposition : If σ and σ' are cohomologous, then $C_c(G,\sigma)$ and $C_c(G,\sigma')$ are isomorphic.

Proof : If $\sigma'(x,y) = \sigma(x,y)c(y)c(xy)^{-1}c(x)$ we can define the isomorphism ϕ from $C_c(G,\sigma')$ to $C_c(G,\sigma)$ which sends f to fc. Indeed for $f,g \in C_c(G)$

$$\begin{aligned}\phi(f) * \phi(g)(x) &= \int f(xy)c(xy)g(y^{-1})c(y^{-1})\,\sigma(xy,y^{-1})d\lambda^{d(x)}(y)\\ &= \int f(xy)g(y^{-1})c(y^{-1})\,\sigma'(xy,y^{-1})c(x)d\lambda^{d(x)}(y)\\ &= \phi(f*g)(x)\end{aligned}$$

$$\begin{aligned}\phi(f)^*(x) &= \overline{\phi(f)(x^{-1})}\;\overline{\sigma(x,x^{-1})} = \overline{f(x^{-1})}\;\overline{c(x^{-1})}\;\overline{\sigma(x,x^{-1})}\\ &= \overline{f(x^{-1})}\;\overline{\sigma'(x,x^{-1})}\;c(x) = \phi(f^*)(x).\end{aligned}$$

Q.E.D.

1.3. Definition : A representation of $C_c(G,\sigma)$ on a Hilbert space H is a $*$-homomorphism $L : C_c(G,\sigma) \to \mathcal{B}(H)$ which is continuous when $C_c(G,\sigma)$ has the inductive limit topology and $\mathcal{B}(H)$ the weak operator topology, and is such that the linear span of $\{L(f)\xi\ ,\ f \in C_c(G,\sigma), \xi \in H\}$ is dense in H.

The I-norm introduced by P. Hahn in [45], page 38, will be a convenient estimate for the C^*-norm we wish to define on $C_c(G,\sigma)$. It is worthwhile to notice that this norm is used by numerical analysts, in the case when G is the trivial equivalence relation on a set of n elements (e.g. in interpolation theory). Let us recall its definition - or rather, the definition appropriate to our setting. For $f \in C_c(G)$

$$\|f\|_{I,r} = \sup_{u\in G^0} \int |f|\,d\lambda^u,\quad \|f\|_{I,d} = \sup_{u\in G^0} \int |f|\,d\lambda_u\ ;\ \text{and}\ \|f\|_I = \max(\|f\|_{I,r}, \|f\|_{I,d}).$$

I.4. Proposition :

(i) $\|\cdot\|_I$ is a norm on $C_c(G)$ defining a topology coarser than the inductive limit topology.

(ii) For any $\sigma \in Z^2(G,\mathbb{T})$, $\| \ \|_I$ is a $*$-algebra norm on $C_c(G,\sigma)$.

Proof :

(i) It suffices to look at $\|f\|_{I,r}$. It is routine to check that it is a norm. Suppose that $f_n \to 0$ in $C_c(G)$. Then, because of the continuity of the map $\lambda : C_c(G) \to C_c(G^0)$ which sends f to $\int f d\lambda^u$, it follows that $\lambda(|f_n|)$ tends to zero in $C_c(G^0)$ and a fortiori in the space $C_0(G^U)$ of continuous functions tending to 0 at infinity, equipped with the supnorm.

(ii) To show that $\|f * g\|_I \le \|f\|_I \ \|g\|_I$, it suffices to consider $\| \ \|_{I,r}$. Then for $f,g \in C_c(G)$,

$$\begin{aligned}\int |f * g| d\lambda^u &\le \iint |f(y)| \ \ |g(y^{-1}x)| d\lambda^{r(x)}(y) d\lambda^u(x), \text{ (because} |\sigma| = 1)\\ &\le \int |f(y)| \int |g(y^{-1}x)| d\lambda^{r(y)}(x) \ d\lambda^u(y)\\ &\le \int |f(y)| \int |g(x)| d\lambda^{d(y)}(x) \ d\lambda^u(y)\\ &\le \sup_v \int |g(x)| d\lambda^v(x) \times \int |f(y)| d\lambda^u(y) \le \|g\|_{I,r} \|f\|_{I,r}.\end{aligned}$$

Finally, by definition, $\|f^*\|_I = \|f\|_I$.

Q.E.D.

1.5. Definition : A representation L of $C_c(G,\sigma)$ will be called bounded if $\|L(f)\| \le \|f\|_I$ for all $f \in C_c(G,\sigma)$.

We may define $\|f\| = \sup \|L(f)\|$ where L ranges over all bounded representations of $C_c(G,\sigma)$, and make two comments.

(i) Clearly $\|\cdot\|$ is a C^*-semi-norm. It will be shown soon, by exhibiting sufficiently many bounded representations, that it is a norm.

(ii) For a large class of groupoids (including transformation groups), we will establish in Corollary 1.22 that every representation on a separable Hilbert space is bounded. This is done in a fashion similar to [24], 4.9, page 45 in the case of a transformation group.

The notion of Hilbert bundle (or more precisely Hilbert space bundle) used in the next definition is given in [61] page 264. The base space of such a bundle is a standard measure space and each fiber is a separable Hilbert space.

<u>1.6. Definition :</u> Let σ be a (continuous) 2-cocycle. A <u>σ-representation of G</u> consists of a quasi-invariant measure μ on G^0 and a (σ,G)-Hilbert bundle $\mathcal{K}$ over (G^0,μ). More precisely, there is a map $L : G \to \mathrm{Iso}(\mathcal{K}) = \{$isometries $\phi_{u,v} : \mathcal{K}_v \to \mathcal{K}_u$ where $u, v \in G^0\}$ such that

(i) $L(x)$ sends $\mathcal{K}_{d(x)}$ onto $\mathcal{K}_{r(x)}$ for $x \in G$ and $L(u) = \mathrm{id}\mathcal{K}_u$ for $u \in G^0$;

(ii) $L(x)L(y) = \sigma(x,y)\, L(xy)$ for ν^2 a.e. (x,y), where ν^2 is the induced measure on G^2 ;

(iii) $L(x)^{-1} = \overline{\sigma(x,x^{-1})}$ for ν a.e. x, where ν is the induced measure on G ; and

(iv) $x \mapsto (L(x)\ \ \xi\circ d(x), \eta\circ r(x))$ is measurable for every pair of measurable sections ξ and η.

Two σ-representations $(\mu,\mathcal{K},L)$ and $(\mu',\mathcal{K}',L')$ are equivalent if the measures μ and μ' are equivalent and there exists an isomorphism ϕ of $\mathcal{K}$ onto $\mathcal{K}'$ (in the sense of [61]) which intertwines L and L', that is, such that $L'(x)\ \phi\circ d(x) = \phi\circ r(x)\ L(x)$ for ν a.e. x.

Let $(\mu,\mathcal{K},L)$ be a σ-representation of G. Then $\Gamma_\mu(\mathcal{K})$, or $\Gamma(\mathcal{K})$ when there is no ambiguity about μ, denotes the Hilbert space of square-integrable sections with respect to μ. The modular function of μ is denoted by D and its symmetric induced measure is $\nu_0 = D^{-1/2}\nu$ (see 1.3.4).

<u>1.7. Proposition :</u> Let $(\mu,\mathcal{K},L)$ be of a σ-representation of G.For $\xi,\eta \in \Gamma(\mathcal{K})$ and $f \in C_c(G)$, set

$$(*)\quad (L(f)\xi,\eta) = \int f(x)\,(L(x)\ \xi\circ d(x), \eta\circ r(x))\, d\nu_0(x).$$

(i) This defines a bounded representation of $C_c(G,\sigma)$ on $\Gamma(\mathcal{K})$.

(ii) Two equivalent σ-representations of G give two equivalent representations of $C_c(G,\sigma)$.

Before starting the proof, let us make two remarks.

a. Let $U_p = \{u \in G^0 : \dim \mathcal{K}_u = p\}$ for $p = 1,2,\ldots,\infty$. It is an invariant measurable subset. Let μ_p be the restriction of μ to U_p.
Then $(\mu_p,\mathcal{K},L)$ is a σ-representation of G, which defines by (*) an operator $L_p(f)$ on $\Gamma_{\mu_p}(\mathcal{K})$; the operator $L(f)$ is the direct sum of the $L_p(f)$'s. Therefore, it is

sufficient to consider the case when dim $\mathcal{K}_u$ is constant. Then $\mathcal{K}$ is isomorphic to a constant Hilbert bundle with fiber K and $\Gamma_\mu(\mathcal{K})$ is isomorphic to the space $L^2(G^0,\mu,K)$ of square-integrable K-valued functions on (G^0,μ). Moreover, for $f \in C_c(G)$ and $\xi \in L^2(G^0,\mu,K)$, $L(f)$ is given by

$$L(f)\xi\ (u) = \int f(x)\ L(x)\ \ \xi\circ d(x)\ D^{-1/2}(x)\ \ d\lambda^u(x)\ \mu\ \text{a.e.}$$

where the right hand-side may be interpreted as a weak integral in K.

b. In the case of a group, G^0 is reduced to one element and there is a unique invariant measure class. Therefore, there is no need to mention it. Then, a σ-representation in the sense of 1.6 is a projective representation in the usual sense (e.g. [74] page 100). Assume that $\sigma = 1$. Then, the representation given by ($*$) is the integrated form of the unitary representation L. It is not the usual expression since our definition of the involution differs from the usual one by the absence of the modular function. To get its usual expression, it suffices to use the remark (ii) following 1.12.

Proof of the proposition : By remark a, we may assume that $\mathcal{K}$ is a constant Hilbert bundle and that $\Gamma(\mathcal{K}) = L^2(G^0,\mu,K)$.

(i) Let us check that $L(f)$ is a well-defined bounded operator. The map $x \mapsto f(x)\ (L(x)\ \ \xi\circ d(x),\ \eta\circ r(x))$ is measurable and dominated in absolute value by $|f(x)|\ \ |\xi\circ d(x)|\ \ |\eta\circ r(x)|$. This last function is ν_0-integrable, because $\nu_0 = D^{-1/2}\nu$, $\nu^{-1} = D^{-1}\nu$ and use of the Cauchy-Schwarz inequality yields

$$\int |f(x)|\ \ |\xi\circ d(x)|\ \ |\eta\circ r(x)|\ \ d\nu_0(x)$$
$$\leq \left[\int |f(x)|\ \ |\xi\circ d(x)|^2\ d\nu^{-1}(x)\right]^{1/2}\ \left[\int |f(x)|\ \ |\xi\circ r(x)|^2\ \ d\nu(x)\right]^{1/2}$$
$$\leq \left[\int |f(x)|\ \ d\lambda_u(x)\ \ |\xi(u)|^2\ d\mu(u)\right]^{1/2}\ \left[\int |f(x)|\ \ d\lambda^u(x)\ |\eta(u)|^2\ d\mu(u)\right]^{1/2}$$
$$\leq \|f\|_{I,d}^{1/2}\ \ |\xi|\ \ \|f\|_{I,r}^{1/2}\ |\eta|$$
$$\leq \|f\|_I\ |\xi|\ \ |\eta|.$$

Therefore, $x \mapsto f(x)\ (L(x)\ \ \xi\circ d(x),\ \eta\circ r(x))$ is ν_0-integrable and $L(f)$ is a bounded operator of norm $\|L(f)\| \leq \|f\|_I$. The continuity of the map $L : C_c(G) \to \mathcal{B}(\Gamma(\mathcal{K}))$ follows from the previous line. We have to check that L is a $*$-homomorphism. So, let f,g be in $C_c(G)$. On one hand, $(L(f * g)\xi,\eta)$ is equal to

$$\int\left(\int f(xy)\ g(y^{-1})\ \sigma(xy,y^{-1})d\lambda^{d(x)}(y)\ \ (L(x)\ \xi\circ d(x),\ \eta\circ r(x)\right)\ D^{1/2}(x)\ d\lambda_u(x)d\mu(u)$$

$$= \int f(xy)\, g(y^{-1})\ \ \sigma(xy,y^{-1})\ (L(x)\ \ \xi\circ d(x),\ \eta\circ r(x))\ D^{1/2}(x)\ d\nu^2(x,y).$$

The use of Fubini's theorem is justified since we are integrating locally integrable functions on compact sets. On the other hand, we may write the equation

$L(g)\xi(u) = \int g(y)\, L(y)\ \ \xi\circ d(y)\ D^{-1/2}(y)\ d\lambda^u(y)$ μ a.e., so that $(L(f)L(g)\xi,\eta)$ is equal to $\iint f(x)g(y)\ (L(x)(L(y)\ \ \xi\circ d(y),\ \eta\circ r(x))\ D^{-1/2}(y)d\lambda^{d(x)}(y)\ d\lambda_u(x)d\mu(u)$. Setting $(x,y)\ \to (xy,y^{-1})$ and using a result in the proof of 1.3.3, we obtain $\int f(xy)g(y^{-1})\ \sigma(xy,y^{-1})\ (L(x)\ \ \xi\circ r(y),\ \ \eta\circ r(x))\ D^{1/2}(x)\ d\nu^2(x,y)$. This shows that $L(f * g) = L(f)L(g)$. Next, for $f \in C_c(G)$,

$$(L(f^*)\xi,\eta) = \int \overline{f(x^{-1})}\ \overline{\sigma(x,x^{-1})}\ (L(x)\ \ \xi\circ d(x),\ \eta\circ r(x))\ d\nu_0(x)$$

$$= \int \overline{f(x)}\ \overline{\sigma(x^{-1},x)}\ (L(x^{-1})\ \ \xi\circ r(x),\ \eta\circ d(x))\ d\nu_0(x),$$

(by symmetry of ν_0).

$$= \int \overline{f(x)}\ \overline{\sigma(x^{-1},x)\ \sigma(x,x^{-1})}\ (\xi\circ r(x),\ L(x)\eta\circ d(x))\ d\nu_0(x)$$

$$= \int \overline{f(x)}\ (\xi\circ r(x),\ L(x)\ \ \eta\circ d(x))\ d\nu_0(x)$$

$$= \ (\xi,L(f)\eta).$$

Finally, the representation L is non-degenerate. Indeed, let η be a vector of $\Gamma(\mathcal{K})$ such that $(L(f)\xi,\eta) = 0$ for every $f \in C_c(G)$ and every $\xi \in \Gamma(\mathcal{K})$. Then, $(L(x)$ $\xi\circ d(x),\ \eta\circ r(x)) = 0$ for ν a.e. x. Choosing a countable total set in K, one sees that $\eta\circ r(x) = 0$ for ν a.e. x.Hence $\eta(u) = 0$ for μ a.e. u.

(ii) Let $(\mu,\mathcal{K},L)$ and $(\mu',\mathcal{K}',L')$ be two equivalent σ-representations. Let g be a positive locally integrable function on (G^0,μ) such that $\mu' = g\mu$ and ϕ an isomorphism of $\mathcal{K}$ onto $\mathcal{K}'$ intertwining L and L'. For $\xi \in \Gamma_\mu(\mathcal{K})$ define $\xi' \in \Gamma_{\mu'}(\mathcal{K}')$ by the formula $\xi'(u) = g^{-1/2}(u)\ \ \phi(u)\ \xi(u)$. Then, the map $\xi \to \xi'$ is an isometry, also denoted ϕ, of $\Gamma_\mu(\mathcal{K})$ onto $\Gamma_{\mu'}(\mathcal{K}')$ which intertwines the integrated representations L and L'. For,

$L(f)\xi(u) = \int f(x)\ L(x)\ \ \xi\circ d(x)\ D^{-1/2}(x)\ d\lambda^u(x)$

$L'(f)\xi'(u) = \int f(x)\ L'(x)\ \xi'\circ d(x)\ D'^{-1/2}(x)\ d\lambda^u(x)$ and

$D'(x) = g\circ r(x)\ D(x)\ (g\circ d(x))^{-1}$ ν a.e. (1.3.3). Thus $(L'(f)\phi\xi)(u)$ is equal to $\int f(x)\ L'(x)(g\circ d(x)^{-1/2}\ \ \phi\circ d(x)\ \ \xi\circ d(x)(g\circ r(x))^{-1/2}\ D^{-1/2}(x)(g\circ d(x))^{1/2}d\lambda^u(x)$

$$= \int f(x)\, g(u)^{-1/2}\, \phi(u)\, L(x)\ \ \xi\circ d(x)\, D^{-1/2}(x)\, d\lambda^u(x)$$
$$= \ (\phi L(f)\xi)\ \ (u).$$

Q.E.D.

It is now easy to construct a faithful family of bounded representations of $C_c(G,\sigma)$, namely the regular representations. As in the case of a group, they play an essential role in the theory of groupoids. They have been defined and thoroughly studied by P. Hahn in [45], where it is pointed out that they have been long-time favorites to produce von Neumann algebras (by the so-called group-measure space construction).

Let σ be a 2-cocycle and μ a quasi-invariant measure. Consider the measurable field of Hilbert space $\{L^2(G,\lambda^u),\ u \in G^o\}$ with square integrable sections $\int^{\oplus} L^2(G,\lambda^u)\, d\mu(u) = L^2(G,\nu)$. For $x \in G$, define $L(x)$ mapping $L^2(G,\lambda^{d(x)})$ to $L^2(G,\lambda^{r(x)})$ by $L(x)\xi(y) = \sigma(x,x^{-1}y)\xi(x^{-1}y)$. This yields a σ-representation of G (cf. example 3.11 of [45]) :

$$\begin{aligned}(L(x)L(y)\xi)(z) &= \sigma(x,x^{-1}z)(L(y)\xi)(x^{-1}z)\\ &= \sigma(x,x^{-1}z)\sigma(y,y^{-1}x^{-1}z)\xi(y^{-1}x^{-1}z)\\ &= \sigma(x,y)\sigma(xy,y^{-1}x^{-1}z)\xi((xy)^{-1}z)\\ &= \sigma(x,y)\ L(xy)\xi(z).\end{aligned}$$

The argument 1.1 shows that the function $(L(x)\ \ \xi\circ d(x),\ \ \eta\circ r(x)) = \int \sigma(x,x^{-1}y)\xi(x^{-1}y)\ \overline{\eta(y)}\ d\lambda^{r(x)}(y)$ is a continuous function of x for $\xi,\eta \in C_c(G)$. Since any vector in $L^2(G,\nu)$ is a pointwise limit of a sequence in $C_c(G)$, this function is measurable when ξ and η are in $L^2(G,\nu)$.

<u>1.8. Definition :</u> The above σ-representation of G will be called the <u>σ-regular representation of G on μ</u>. Its integrated form is the <u>regular representation on μ</u> of $C_c(G,\sigma)$.

It is a basic fact ([45], theorem 2.15) that the regular representation on μ is the left representation of a left Hilbert algebra. We reproduce it in our context. The main ingredient of the proof, which is the construction of a left approximate identity, will be needed in other places.

1.9. Proposition : The algebra $C_c(G,\sigma)$ has a left approximate identity (for the inductive limit topology).

Proof : Let us say that a subset A of G is d-relatively compact if $A \cap d^{-1}(K)$ is relatively compact for any compact subset K of G^0. Then, if L is relatively compact, $AL = (A \cap d^{-1}(r(L)))L$ is also relatively compact. Let us show that G^0 has a fundamental system of d-relatively compact neighborhoods. Let V be an open neighborhood of G^0 and (K_i) a locally finite relatively compact open cover of G^0 (in G). There exists a relatively compact open set U_i in G such that $K_i \subset U_i \subset V \cap d^{-1}(K_i)$. Then $U = \cup U_i$ is an open neighborhood of G^0 contained in V and is d-relatively compact. Indeed, since any compact subset K of G^0 meets only a finite number of K_i's, $U \cap d^{-1}(K)$ is contained in a finite union of U_i's. Let (U_α) be such a fundamental system, with $U_\alpha \subset U_1$ for every α and let (K_α) be a net of compact subsets of G^0 increasing to G^0. We can find non-negative $g_\alpha \in C_c(G)$, positive on K_α and with support contained in U_α and non-negative $h_\alpha \in C_c(G^0)$ such that $h_\alpha(u) = (\int g_\alpha d\lambda^u)^{-1}$ for $u \in K_\alpha$. Let us define $f_\alpha(x) = h_\alpha \circ r(x)\, g_\alpha(x)$. Then, $f_\alpha \in C_c(G)$, supp $f_\alpha \subset U_\alpha$ and $\lambda(f_\alpha) = 1$ on K_α. We claim that (f_α) is a left approximate identity. Let $f \in C_c(G)$ with K = suppf. Then supp$(f_\alpha * f)$ and suppf are contained in the compact set $L = \overline{U_1 K}$. If $\varepsilon > 0$ is given, the using the compactness of L and the continuity of f,σ and the product, one may find α_0 such that for $\sigma > \alpha_0$ and every $(x,y) \in L \times U_\alpha \cap G^2$, $|f(y^{-1}x) - f(x)| \leq \varepsilon$ and $|\sigma(y,y^{-1}x) - 1| \leq \varepsilon$ while $r(L) \subset K_\alpha$. It follows that

$$f_\alpha * f(x) - f(x) = \int f_\alpha(y)[f(y^{-1}x) - f(x)]\sigma(y,y^{-1}x)d\lambda^{r(x)}(y)$$

$$+ \quad f(x) \cdot \int f_\alpha(y)[\sigma(y,y^{-1}x) - 1] \; d\lambda^{r(x)}(y), \text{ and}$$

$$|f_\alpha * f(x) - f(x)| \leq \varepsilon + \sup_y |f(y)| \, \varepsilon \text{ for } x \in L.$$

Q.E.D.

If (f_α) is a left approximate identity, (f_α^*) is a right approximate identity. I have not been able to prove the existence of a two-sided approximate identity for $C_c(G)$ except in particular cases (r-discrete groupoid and transformation groups).

The definition of a generalized Hilbert algebra, used in the next proposition, can be found in [73] , pages 5 and 6.

1.10. Proposition : (Cf. theorem 2.15 of [45]). Let σ be a 2-cocycle and μ a quasi-invariant measure. Then

(i) $C_c(G,\sigma)$ with the inner product of $L^2(G,\nu^{-1})$ is a generalized Hilbert algebra ; and

(ii) its left representation is equivalent to the regular representation on μ.

Proof : Let us check the axioms of [73].

(I) For f,g and h in $C_c(G,\sigma)$,

$$(g, f^* * h) = \int g(y)\, f^* * h(y)\, d\nu^{-1}(y)$$
$$= \iiint g(y)\, f^*(yx)\, h(x^{-1})\, \sigma(yx,x^{-1})\, d\lambda^{d(y)}(x)\, d\lambda_u(y)\, d\mu(u)$$
$$= \iiint g(y)\, f^*(yx)\, h(x^{-1})\, \sigma(yx,x^{-1})\, d\lambda_{r(x)}(y)\, d\lambda^u(x)\, d\mu(u)$$

(the use of Fubini's theorem is justified because the function $(x,y) \mapsto f(y)\, f^*(yx)\, h(x^{-1})\sigma(yx,x^{-1})$, defined on $G^u \times G_u$, is continuous with compact support)

$$= \iiint g(y)\, f^*(yx^{-1})\, h(x)\, \sigma(yx^{-1},x)\, d\lambda_{d(x)}(y)\, d\lambda_u(x)\, d\mu(u)$$
$$= \iiint f(y^{-1})\, f^*(y^{-1}x^{-1})\, h(x)\, \sigma(y^{-1}x^{-1},x)\, d\lambda^{d(x)}(y)\, d\lambda_u(x)\, d\mu(u)$$
$$= \iiint g(y^{-1})\, f(xy)\sigma(y^{-1}x^{-1},xy)\, h(x)\, \sigma(y^{-1}x^{-1},x) d\lambda^{d(x)}(y)\, d\lambda_u(x)\, d\mu(u)$$
$$= \iiint f(xy)\, g(y^{-1})\, \sigma(xy,y^{-1}) d\lambda^{d(x)}(y)\, h(x)\, d\lambda_u(x)\, d\mu(u)$$
$$= (f * g, h).$$

(II) For every $f \in C_c(G)$, $g \mapsto f * g$ is continuous. In fact, this operator has norm $\leq \|f\|_I$, as it can be seen directly or deduced from (ii).

(III) Since $C_c(G,\sigma)$ has a left approximate identity, the set $\{f * g : f, g \in C_c(G)\}$ is dense in $C_c(G)$ with the inductive limit topology and a fortiori with the $L^2(G,\nu^{-1})$ topology.

(IX) We have to show that the involution, as a real linear operator, is closable. Suppose that $f_n \to 0$ and $f_n^* \to g$. Then $\int |f_n|^2 d\nu^{-1} \to 0$ and $\int |f_n^*(x) - g(x)|^2\, d\nu^{-1}(x) = \int |f_n(x) - g^*(x)|^2\, d\nu(x) \to 0$. Thus there is a subsequence f_{n_k} such that $f_{n_k} \to 0$ ν^{-1} a.e. and $f_{n_k} \to g^*$ ν a.e. Since ν and ν^{-1} are equivalent, $g^* = 0$ ν a.e., hence $g = 0$.

(ii) Let us call L' the left representation on $L^2(G,\nu^{-1})$, $L'(f)g = f * g$, and L the regular representation on μ acting on $L^2(G,\nu)$. The isometry from $L^2(G,\nu)$ onto

$L^2(G,\nu^{-1})$ which sends ξ into $\xi' = D^{1/2}\xi$ implements their equivalence. For $\xi,\eta \in L^2(G,\nu)$ and $f \in C_c(G,\nu)$,

$$
\begin{aligned}
(L'(f)\xi', \eta') &= \int f * \xi'(y)\ \bar{\eta}'(y)\ d\nu^{-1}(y) \\
&= \iiint f(x)\ \xi'(x^{-1}y)\ \sigma(x,x^{-1}y)\ \overline{\eta'(y)}\ d\lambda^{r(y)}(x)\ d\lambda_u(y)\ d\mu(u) \\
&= \iiint f(x)\ D^{1/2}(x^{-1}y)\ \xi(x^{-1}y)\ \sigma(x,x^{-1}y)\ D^{1/2}(y)\ \overline{\eta(y)} \\
&\qquad d\lambda^{r(y)}(x)\ d\lambda_u(y)\ d\mu(u) \\
&= \iiint f(x)\ \xi(x^{-1}y)\ \sigma(x,x^{-1}y)\ \overline{\eta(y)}\ D^{-1/2}(x)\ d\lambda^{r(y)}(x) d\lambda^u(y)\ d\mu(u) \\
&= \iiint f(x)\ \xi(x^{-1}y)\ \sigma(x,x^{-1}y)\ \overline{\eta(y)}\ d\lambda^{r(x)}(y)\ D^{-1/2}(x)\ d\lambda^u(x)\ d\mu(u) \\
&= \int f(x)\ (L(x)\ \xi\circ d(x),\ \eta\circ r(x))\ d\nu_0(x) \\
&= (L(f)\xi,\eta).
\end{aligned}
$$

Q.E.D.

By looking at the polar decomposition of the involution, we obtain the usual ingredients of the Tomita-Takesaki theory : the modular involution $J : L^2(G,\nu^{-1}) \to L^2(G,\nu^{-1})$ is given by $J\xi(x) = D^{1/2}(x)\ \overline{\xi(x^{-1})}\ \overline{\sigma(x,x^{-1})}$, and the modular operator Δ is defined on $L^2(G,\nu) \cap L^2(G,\nu^{-1})$ by $\Delta\xi(x) = D(x)\ \xi(x)$.

<u>1.11. Proposition :</u> $C_c(G,\sigma)$ has a faithful family of bounded representations, consisting of regular representations.

<u>Proof :</u> Let μ be a quasi-invariant measure with induced measure ν and let L be the regular representation of $C_c(G,\sigma)$ on μ. The kernel of L is $\{f \in C_c(G,\sigma) : f$ vanishes on $\mathrm{supp}\,\nu\}$. For, if f vanishes on $\mathrm{supp}\,\nu$, the formula 1.7 (*) shows that $L(f) = 0$; while conversely, if $f * g = 0$ in $L^2(G,\nu^{-1})$ for any $g \in C_c(G)$, then using a right approximate identity, one sees that $f = 0$ in $L^2(G,\nu^{-1})$, so that f vanishes on $\mathrm{supp}\,\nu$. To conclude, we observe that G^0 has a faithful family of quasi-invariant measures, the transitive measures (definition 1.3.9).

Q.E.D.

<u>1.12. Definition :</u> Let σ be a 2-cocycle. The C^*-algebra $C^*(G,\sigma)$ is the completion of $C_c(G,\sigma)$ for the C^*-norm defined in 1.5. It is called the <u>σ-C^*algebra of the groupoid G.</u> The <u>C^*-algebra of G</u> is $C^*(G) = C^*(G,1)$.

Remarks :

(i) It results from 1.2 that cohomologous 2-cocycles give isomorphic C^*-algebras.

(ii) In the case of a group or a transformation group, our definition does not quite agree with the usual one (eg. [24] p. 35) because of the absence of a modular function in the involution. However, the C^*-algebras are isomorphic. To see this, let $G = U \times S$. We denote by $C_c^*(G)$ the $*$-algebra of 1.1 and by $C_{c*}(G)$ the $*$-algebra of [24]. The involution for the latter is $f_*(u,s) = f^*(u,s)\,\Delta(s^{-1})$, where Δ is the modular function of the group S. Then the map from $C_c^*(G)$ to $C_{c*}(G)$ sending f to $f'(u,s) = f(u,s)\Delta^{-1/2}(s)$ is a $*$-isomorphism. It extends to an isomorphism of $C^*(G)$ onto $C_*(G)$.

(iii) If G is second countable, then $C_c(G)$ with the inductive limit topology is separable, therefore $C^*(G,\sigma)$ is separable.

If h is a bounded continuous function on G^0 and $f \in C_c(G)$, we define

$hf(x) = h\circ r(x)\ f(x)$, and

$fh(x) = f(x)\ h\circ d(x)$.

Then, hf and fh $\in C_c(G)$ and the following relations hold in the $*$-algebra $C_c(G,\sigma)$. For every f, g $\in C_c(G,\sigma)$,

$f * hg = fh * g$,

$h(f * g) = hf * g$, and

$(hf)^* = f^*h^*$ where $h^*(u) = \overline{h(u)}$.

For example,

$$
\begin{aligned}
f * hg(x) &= \int f(xy)\ hg(y^{-1})\ \sigma(xy,y^{-1})\,d\lambda^{d(x)}(y)\\
&= \int f(xy)\ h\circ d(y)\ g(y^{-1})\ \sigma(xy,y^{-1})\ d\lambda^{d(x)}(y)\\
&= \int fh(xy)\ g(y^{-1})\ \sigma(xy,y^{-1})\ d\lambda^{d(x)}(y)\\
&= fh * g\ (x)
\end{aligned}
$$

In other words, h acts on $C_c(G,\sigma)$ as a double centralizer (cf. [47]). Moreover it acts continuously with respect to the inductive limit topology.

1.13. Lemma : If L is a representation of $C_c(G,\sigma)$, there exists a unique representation

M of $C_c(G^0)$ such that for every $h \in C_c(G^0)$ and every $f \in C_c(G,\sigma)$, $L(hf) = M(h)L(f)$ and $L(fh) = L(f)M(h)$.

Proof : Let H be the space of the representation L and H_0 the linear span of $\{L(f)\xi : f \in C_c(G,\sigma),\ \xi \in H\}$. Let us try to define M(h) on H_0 by $M(h)\ (\sum_1^n L(f_i)\xi_i) = \sum_1^n L(hf_i)\xi_i$, just as in [47], page 317. To show that M(h) is well defined, it suffices to prove

$$\sum_1^n L(f_i)\xi_i = 0 \Longrightarrow \sum_1^n L(hf_i)\xi_i = 0.$$

Let (f_α) be a left approximate identity for $C_c(G,\sigma)$. Then

$$\sum_1^n L(hf_i)\xi_i = \lim \sum_1^n L(h(f_\alpha) * f_i))\xi_i = \lim \sum_1^n L(hf_\alpha * f_i)\xi_i$$
$$= \lim L(hf_\alpha) \sum_1^n L(f_i)\xi_i = 0 .$$

Moreover, M(h) satisfies

$$(M(h)L(f)\xi, L(g)\eta) = (L(hf)\xi, L(g)\eta) = (\xi, L(hf)^* L(g)\eta)$$
$$= (\xi, L(f^* * h^* g)\eta) = (\xi, L(f^*)L(h^* g)\eta)$$
$$= (L(f)\xi, L(h^* g)\eta) = (L(f)\xi, M(h^*)L(g)\eta).$$

To show that M(h) is a bounded operator, one uses as in [24], page 41, the relation

$$(hg)^* * (hf) + (kg)^* * (kf) = \|h\|^2 g^* * f$$

valid for every $h \in C_c(G^0)$, $f, g \in C_c(G,\sigma)$, where $k(u) = (\|h\|^2 - |h(u)|^2)^{1/2}$. Then

$$\|M(h) \sum_1^n L(f_i)\xi_i\|^2 = \sum_{i,j} (L(hf_i)\xi_i, L(hf_j)\xi_j)$$
$$= \sum_{i,j} (L((hf_j)^* * (hf_i))\xi_i, \xi_j)$$
$$= \|h\|^2 \sum_{i,j} (L(f_j^* * f_i)\xi_i, \xi_j) - \sum_{i,j} (L(kf_j)^* * (kf_i))\xi_i, \xi_j)$$
$$= \|h\|^2 \|\sum_i L(f_i)\xi_i\|^2 - \|\sum_i L(kf_i)\xi_i\|^2$$
$$\leq \|h\|^2 \|\sum_i L(f_i)\xi_i\|^2.$$

Therefore M(h) extends to a bounded operator on H. It is then routine to check that M is a representation of the $*$-algebra $C_c(G^0)$ and that $L(fh) = L(f)M(h)$. Q.E.D.

1.14. Proposition : The C^*-algebra $C^*(G^0)$ is a subalgebra of the multiplier algebra of $C^*(G,\sigma)$.

Proof : The action of $C_c(G^0)$ as double centralizers of $C_c(G,\sigma)$ extends to $C^*(G,\sigma)$, because for every bounded representation L of $C_c(G,\sigma)$,

$$\|L(hf)\| \leq \|M(h)\| \ \|L(f)\| \leq \|h\| \ \|f\| \quad \text{, hence}$$

$$\|hf\| \leq \|h\| \ \|f\| .$$

This gives a $*$-homomorphism of $C^*(G^0)$ into the multiplier algebra of $C^*(G,\sigma)$ which is visibly one-to-one.

Q.E.D.

The notion of generalized expectation used in the next proposition was introduced by M. Rieffel in [63] (definition 4.12) in a context close to this one. We shall look at it again in the second section.

1.15. Proposition : The restriction map $C_c(G,\sigma) \to C_c(G^0)$ is a generalized expectation. It is smooth and faithful.

The proof will be given in a more general situation in the second section (2.9).

Remark : If G is r-discrete, $C^*(G^0)$ is a subalgebra of $C^*(G,\sigma)$ and the restriction map of $C_c(G)$ onto $C_c(G^0)$ extends to an expectation of $C^*(G,\sigma)$ onto $C^*(G^0)$. In this case, $C^*(G,\sigma)$ has a unit iff G^0 is compact.

It will be convenient in the following discussion to enlarge the class of functions on G.

1.16. Proposition : Let B(G) denote the set of bounded Borel functions on G with compact support. With convolution and involution defined as in 1.1, B(G) can be made into a $*$-algebra, denoted $B(G,\sigma)$.

The proof is similar to 1.1. One can also use 1.1 and the fact that any function in B(G) is a bounded pointwise limit of a sequence of functions in $C_c(G)$.

Moreover, we may define the following notion of convergence in $B(G,\sigma)$: a sequence (f_n) in $B(G,\sigma)$ converges to $f \in B(G,\sigma)$ iff $f_n(x) \to f(x)$ for every $x \in G$ and there exists $h \in B(G)$ such that $|f_n| \leq h$ and $|f| \leq h$. Then $f_n \to f$, $g_n \to g \Longrightarrow f_n * g_n \to f * g$ and $f_n^* \to f^*$.

Let us define a representation of $B(G,\sigma)$ as a $*$-homomorphism $L : B(G,\sigma) \to \mathcal{B}(H)$, where H is a Hilbert space, which is continuous in the sense $f_n \to f \Longrightarrow (L(f_n)\xi,\eta) \to (L(f)\xi,\eta)$ for any $\xi,\eta \in H$, and is such that the linear span of $\{L(f)\xi,\ f \in B(G,\sigma),\ \xi \in H\}$ is dense in H.

<u>1.17. Lemma :</u> Every representation of $C_c(G,\sigma)$ extends to a representation of $B(G,\sigma)$.

<u>Proof :</u> Suppose that $f \in B(G)$. There exists a sequence (f_n) in $C_c(G)$ converging to f in $B(G,\sigma)$. By Lebesgue's dominated convergence theorem, for every ξ,η in the space H of the representation L, f is integrable with respect to the measure $(L(\ )\xi,\nu)$ and $(L(f_n)\xi,\eta) \to (L(f)\xi,\eta)$. By the uniform boundedness theorem, $L(f)$ is a bounded operator. To show that L is a $*$-homomorphism, we use again an approximation argument. The continuity of L results from Lebesgue's dominated convergence theorem.

Q.E.D

The next goal is to realize the inverse semi-group $\mathcal{G}_b$ of non-singular Borel G-sets (1.3.26) as an inverse semi-group of partial isometries. For $S \in \mathcal{G}_b$ and $f \in B(G)$, we define

$$sf(x) = \delta^{1/2}(r(x),s)f(s^{-1}x)\sigma(s,s^{-1}x) \text{ if } x \in r^{-1}(r(S)),$$
$$= 0 \text{ if } x \notin r^{-1}(r(S)) \ ;$$
$$fs(x) = \delta^{1/2}(d(x),s^{-1})f(xs^{-1})\sigma(xs^{-1},s) \text{ if } x \in d^{-1}(d(S)),$$
$$= 0 \text{ if } x \notin d^{-1}(d(S)) \ ; \text{ and}$$
$$s^*f = \overline{\tilde{\sigma}(s^{-1},s)}\ (s^{-1}f),$$

where $\delta(\cdot,s)$ denotes the vertical Radon-Nikodym derivative of S. The notations have been defined in 1.1.11 and 1.1.18. For convenience, $\sigma(s,s^{-1}x)$ is written instead of $\sigma(sr(s^{-1}x),s^{-1}x)$. In accordance with 1.1.18,

$$\tilde{\sigma}(s,t)\ (u) = \sigma(us,(u\cdot s)t)\ .$$

Also for a bounded Borel function h on G^0 and $f \in B(G)$, $hf(x) = h\circ r(x)\ f(x)$. We note that sf, fs and s^*f are functions in $B(G)$.

1.18. Lemma : The following relations hold in the $*$-algebra $B(G,\sigma)$:

(i) $s(tf) = \tilde{\sigma}(s,t)\ (st)f$ for $s,t \in \mathcal{G}_b$ and $f \in B(G)$;

(ii) $fs*g = f*sg$ for $f,g \in B(G)$ and $s \in \mathcal{G}_b$;

(iii) $(fs)^* = s^* f^*$ for $f \in B(G)$ and $s \in \mathcal{G}_b$;

(iv) $s(f*g) = sf*g$ for $f,g \in B(G)$ and $s \in \mathcal{G}_b$; and

(v) $f_n \to f \Rightarrow sf_n \to sf$ for f_n, $f \in B(G)$ and $s \in \mathcal{G}_b$.

Proof : The verifications are straightforward computations.

(i)
$$\begin{aligned} s(tf) &= \delta^{1/2}(r(x),s)\ tf(s^{-1}x)\ \sigma(s,s^{-1}x) \text{ for } x \in r^{-1}(r(st)) \\ &= \delta^{1/2}(r(x),s)\ \delta^{1/2}(r(x)\cdot s,t)\ f(t^{-1}s^{-1}x)\ \sigma(t,t^{-1}s^{-1}x)\sigma(s,s^{-1}x) \\ &= \delta^{1/2}(r(x),st)\ f(t^{-1}s^{-1}x)\ \tilde{\sigma}(s,t)\circ r(x)\ \sigma(st,(st)^{-1}(x) \\ &= \sigma(s,t)\ (st)\ f\ (x), \text{ and} \\ &= 0 \text{ for } x \notin r^{-1}(r(st)). \end{aligned}$$

(ii)
$$\begin{aligned} fs*g(x) &= \int fs(y)\ g(y^{-1}x)\ \sigma(y,y^{-1}x)\ d\lambda^{r(x)}(y) \\ &= \int \delta^{1/2}(d(y),s^{-1})f(ys^{-1})g(y^{-1}x)\ \sigma(ys^{-1},s)\ \sigma(y,y^{-1}x)d\lambda^{r(x)}(y). \end{aligned}$$

Changing the variable y into ys, this last expression becomes

$$\begin{aligned} &\int \delta^{1/2}(d(y)\cdot s,s^{-1})\ \delta(d(y),s)\ f(y)\ g(s^{-1}y^{-1}x)\ \sigma(y,s)\sigma(ys,s^{-1}y^{-1}x) \\ &\qquad d\lambda^{r(x)}(y) \\ &= \int \delta^{1/2}(d(y),s)\ f(y)\ g(s^{-1}y^{-1}x)\sigma(y,y^{-1}x)\sigma(s,s^{-1}y^{-1}x)\ d\lambda^{r(x)}(y) \\ &= \int f(y)\ sg(y^{-1}x)\ \sigma(y,y^{-1}x)\ d\lambda^{r(x)}(y) \\ &= f*sg\ (x). \end{aligned}$$

(iii)
$$\begin{aligned} (fs)^*(x) &= \overline{fs}(x^{-1})\ \overline{\sigma}(x,x^{-1}) \\ &= \delta^{1/2}(r(x),s^{-1})\ \overline{f}(x^{-1}s^{-1})\ \overline{\sigma}(x^{-1}s^{-1},s)\ \overline{\sigma}(x,x^{-1}) \\ &= \delta^{1/2}(r(x),s^{-1})\ \overline{f}(x^{-1}s^{-1})\ \overline{\sigma}(x,x^{-1}s^{-1})\ \overline{\sigma}(r(x)s^{-1},(r(x)\cdot s^{-1})s) \\ &= \delta^{1/2}(r(x),s^{-1})\ \overline{f}\ (x^{-1}s^{-1})\ \overline{\sigma}(sx,x^{-1}s^{-1})\sigma(s^{-1},sx)\ \overline{\tilde{\sigma}}(s^{-1},s)\circ r(x) \\ &= \delta^{1/2}(r(x),s^{-1})\ \ f\ (sx)\sigma(s^{-1},sx)\ \overline{\tilde{\sigma}}(s^{-1},s)\circ r\ \ (x) \\ &= \overline{\tilde{\sigma}}(s^{-1},s)\ (s^{-1}f)\ (x). \end{aligned}$$

(iv)
$$\begin{aligned} s(f*g)(x) &= \delta^{1/2}(r(x),s)\ f*g\ (s^{-1}x) \\ &= \int \delta^{1/2}(r(x),s)\ \ f(s^{-1}xy)\ g(y^{-1})\ \sigma(xy,y^{-1})\ d\lambda^{d(x)}(y) \\ &= \int \delta^{1/2}(r(xy),s)\ f(s^{-1}xy)\ g(y^{-1})\ \sigma(xy,y^{-1})\ d\lambda^{d(x)}(y) \\ &= sf*g\ (x)\ . \end{aligned}$$

(v) This is clear, since we assume that the vertical Radon-Nikodym derivative $\delta(\cdot,s)$ is bounded on compact sets.

Q.E.D.

1.20. Lemma : Let L be a representation of $B(G,\sigma)$.

(i) There is a unique representation M of $B(G^0)$ such that $L(hf) = M(h)L(f)$ and $L(fh) = L(f)M(h)$ for every $h \in B(G^0)$ and every $f \in B(G)$.

(ii) There is a unique σ-representation V of the Borel ample semi-group $\mathcal{G}_b$ as an inverse semi-group of partial isometries such that $L(sf) = V(s)L(f)$ and $L(fs) = L(f)V(s)$ for every $s \in \mathcal{G}_b$ and every $f \in B(G)$.

(iii) The following covariance relation between V and M holds : $V(s)\, M(h)\, V(s)^*$ $= M(h^s)$ for every $s \in \mathcal{G}_b$ and every $h \in B(G^0)$ where $h^s(u) = h(u\, s)$ if $u \in r(s)$,
$= 0$ if $u \notin r(s)$.

Proof : We first note that the approximate identity constructed in 1.9, which can be chosen countable since G is second countable, satisfies $e_n * f \to f$ for any $f \in C_c(G,\sigma)$. Let L be a representation of $B(G,\sigma)$ on the Hilbert space H and let H_0 be the linear span of $\{L(f)\xi : f \in C_c(G),\ \xi \in H\}$. We proceed as in 1.13 to define $M(h)$ and $V(s)$ on H_0 :

$$M(h)\ \left(\sum_1^n L(f_i)\,\xi_i\right) = \sum_1^n L(hf_i)\xi_i, \text{ and}$$

$$V(s)\ \left(\sum_1^n L(f_i)\xi_i\right) = \sum_1^n L(sf_i)\xi_i\ .$$

We check as in 1.13 that $M(h)$ and $V(s)$ are well defined.

(i) This is obtained as 1.13.

(ii) It is immediate to check the following relations on H_0 :

$V(s)\, V(t) = M(\tilde{\sigma}(s,t))\, V(st)$

$V(s)^{-1} = M(\overline{\tilde{\sigma}(s,s^{-1})})\, V(s^{-1})$

$V(s)\, V(s)^{-1} = M(\chi_{r(S)})$ and $V(s)^{-1}\, V(s) = M(\chi_{d(S)})$, where χ_A is the characteristic function of A ; and

$V(s)^{-1} \subset V(s)^*$

In particular, $V(s)$ is a partial isometry and it extends to H.

(iii) For $s \in \mathcal{G}_b$, $h \in B(G^0)$ and $f \in B(G)$

$$s\, h\, \bar{\tilde{\sigma}}(s^{-1},s)\, s^{-1} f(x) = \delta^{1/2}(r(x),s)\ \ (h\bar{\tilde{\sigma}}(s^{-1},s))s^{-1}f(s^{-1}x)\sigma(s,s^{-1}x)$$
$$= \delta^{1/2}(r(x),s)\, h(r(x)\cdot s)\, \bar{\tilde{\sigma}}(s^{-1},s)\, \delta^{1/2}(r(x).s,s^{-1})\, f(x)\, \sigma(s^{-1},x)\, \sigma(s,s^{-1}x)$$
$$= h(r(x),s)\, f(x)\, \bar{\tilde{\sigma}}(s^{-1},s)\, \sigma(s,s^{-1})$$
$$= (h^s f)\ (x).$$

Therefore,

$$V(s)\, M(h)\, V(s)^* \ L(f) = V(s)\, M(h)\, M(\bar{\tilde{\sigma}}(s^{-1}s))\ \ V(s^{-1})\, L(f)$$
$$= L(sh\, \bar{\tilde{\sigma}}(s^{-1},s)s^{-1}f)$$
$$= L(h^s f) = M(h^s)\ \ L(f).$$

Q.E.D.

We have seen(1.7) that bounded representations of $C_c(G,\sigma)$ could be obtained by integrating σ-representations of G. The correspondence between the unitary representations of a group and the representations of its C^*-algebra is well known and justifies a large part of the theory of C^*-algebras. The generalization of this result which we give in the case of groupoids has a more limited scope. We only consider groupoids which are second countable and representations on separable Hilbert spaces. Moreover, we need an additional assumption on the groupoid, namely, it should admit sufficiently many non-singular G-sets (definition (1.3.27). This assumption is satisfied in the case of a transformation group and of an r-discrete groupoid. I do not have any example where it is not satisfied. The case of a transformation group is not new (e.g. [74], theorem 9.11, page 73). However, the proof usually given uses the standard Borel structure of the group and seems to fail in the case of a groupoid. Instead, we will use part of a theorem of P.Hahn ([43], theorem 5.4, page 106), which will be reproduced below as part of the proof of 1.21 since it has not yet appeared in print.

1.21. Theorem : Let G be a second countable locally compact groupoid with Haar system and with sufficiently many non-singular Borel G-sets and σ a continuous 2-cocycle in $Z^2(G,\mathbb{T})$. Then, every representation of $C_c(G,\sigma)$ on a separable Hilbert space is the integrated form of a σ-representation of G.

Proof : We will only consider factor representations. The general case is then obtained by direct integral decomposition and requires the definition of a direct integral of a family of σ-representations of G. Since this theorem will only be used in the case of factor representations, we omit the general case here. Let L be a factor representation of $C_c(G,\sigma)$ on the separable Hilbert space H. We use 1.17 to extend it to a representation of $B(G,\sigma)$ and 1.20 to obtain the representation M of $B(G^0)$ and the σ-representation V of $\mathcal{G}_b$ such that $L(hf) = M(h)L(f)$ and $L(sf) = V(s)L(f)$. It results from multiplicity theory that there exists a probability measure μ on G^0 and a Hilbert bundle $\mathcal{K}$ over (G^0,μ) such that M is unitarily equivalent to multiplication on the Hilbert space $\Gamma(\mathcal{K})$ of square-integrable sections. From now on, we assume that $H = \Gamma(\mathcal{K})$ and that M is multiplication.

a. Our first task is to show that the measure μ is quasi-invariant. Let ν be its induced measure. We show that for $f \in B(G)$, if $f = 0$ ν a.e., then $L(f) = 0$. Let $E = \{u \in G^0 \ : \ \int |f| d\lambda^u > 0\}$. By assumption, $M(\chi_E) = 0$. If $x \notin r^{-1}(E)$, then for every $g \in B(G)$,

$$f * g(x) = \int f(y)\, g(y^{-1})\, \sigma(y,y^{-1}x) d\lambda^{r(x)}(y) = 0 \text{ and therefore}$$

$$f * g = \chi_E(f * g). \text{ Then}$$

$$L(f)L(g) = L(f * g) = M(\chi_E)\, L(f * g) = 0.$$

Since L is non degenerate, $L(f) = 0$. Thus, for $f \in B(G)$, $L(f)$ depends only on the ν-class of f. To show that μ is quasi-invariant, we pick a Borel set A in G of positive ν-measure and we show that it has positive ν^{-1}-measure. We may assume that for every $u \in r(A)$ and any open set V in G such that $V \cap A^u \neq \emptyset$, $\lambda^u(V \cap A) > 0$. Since G has sufficiently many non-singular Borel G-sets, there exists a non-singular Borel G-set S of positive μ-measure which is contained in A. We can construct a sequence (U_n,e_n) where U_n is a Borel set contained in A and e_n a non-negative function in $B(G)$ vanishing outside U_n such that $\int e_n \, d\lambda^u = 1$ for $u \in r(A)$ and every n and (U_n) shrinks to S in the sense that every neighborhood of S contains U_n for n sufficiently large. Let $f_n(y) = \delta^{1/2}(r(y),s)\, e_n(y)$ for $y \in r^{-1}(r(S))$, 0 otherwise. Then, for every $f \in C_c(G)$,

$$\begin{aligned} f_n * f(x) &= \int \delta^{1/2}(r(y),s)\, e_n(y)\, f(y^{-1}x)\, \sigma(y,y^{-1}x) d\lambda^{r(x)}(y) \\ &= \int \delta^{1/2}(r(x),s)\, e_n(y)\, f(y^{-1}x)\, \sigma(y,y^{-1}x) d\lambda^{r(x)}(y). \end{aligned}$$

Hence, for every x,

$$f_n * f(x) \to \delta^{1/2}(r(x),s)\, f(s^{-1}x)\quad \sigma(s,s^{-1}x) = sf(x)$$

Moreover $|f_n * f|(x) \le |sf|(x)$. Therefore, $L(f_n^0)L(f) = L(f_n * f) \to L(sf) = V(s)L(f)$ in the weak operator topology. If A had zero ν^{-1}- measure, then since supp $f_n^* \subset A^{-1}$, we would have $f_n^* = 0$ ν a.e. and $L(f_n^*)^* = L(f_n^*) = 0$, hence $L(f_n) = 0$. We would conclude that $V(s)L(f) = 0$ for every $f \in C_c(G)$, hence $V(s) = 0$. However this would contradict the fact that $V(s)V(s) = M(\chi_{r(S)}) > 0$.

b. Let us show next that for each non-singular Borel G-set S, the partial isometry V(s) on $\Gamma(\mathcal{K})$ is of the form

$$V(s)\xi\,(u) = \Delta^{1/2}(us,s)\; c(u,s)\; \xi(u\cdot s)\quad \text{for } u \in r(S)$$

$$= 0 \qquad\qquad\qquad\qquad \text{for } u \notin r(S)$$

where $\Delta(\cdot,s)$ is the horizontal Radon-Nikodym derivative of S (1.3.18) and c(u,s), defined for μ a.e. $u \in r(S)$ is an isometry of $\mathcal{K}_{u\cdot s}$ onto $\mathcal{K}_u$. This follows directly from a result of Guichardet ([38], proposition 1, page 82) which we recall here :
Let $\mathcal{K}$ and $\mathcal{L}$ be two Hilbert bundles over the standard measure spaces (X,α) and (Y,β) respectively, ϕ an isomorphism of (X,α) onto (Y,β) and U an isometry of $\Gamma(\mathcal{K})$ onto $\Gamma(\mathcal{L})$ satisfying

$$UM(h)\,U^{-1} = M(h\circ\phi^{-1}) \text{ for every } h \in L^\infty(X,\alpha),$$

where M denotes the multiplication operator. Then, there exist isometries u(y) from $\mathcal{K}_{\phi^{-1}(y)}$ onto $\mathcal{L}_y$ defined for β a.e. y such that for every $\xi \in \Gamma(\mathcal{K})$,

$$U\xi\,(y) = r^{1/2}(y)\; u(y)\; \xi(\phi^{-1}(y))\ \beta \text{ a.e.}$$

where $r = \frac{d\phi\alpha}{d\beta}$ is the Radon-Nikodym derivative of $\phi\alpha$ with respect to β.

c. Next, we show that the set of constant multiplicity $A = \{u \in G^0 : \dim \mathcal{K}_u = p\}$ for $p = 1,2,\dots,\infty$, of the Hilbert bundle $\mathcal{K}$ is almost invariant (definition 1.3.5). If A were not almost invariant, there would be a Borel set B of positive ν-measure such that for every $x \in B$, $r(x) \in A$ and $d(x) \notin A$. By assumption, B contains a non-singular Borel G-set S such that $\mu(r(S)) > 0$. However for μ a.e. $u \in r(S)$, there is an isometry c(u,s) from $\mathcal{K}_{u\cdot s}$ onto $\mathcal{K}_u$, hence $u\cdot s \in A$. This is a contradiction.

d. We show that for a Borel set B in G^0, the projection $M(\chi_B)$ is in the commutant of L iff B is almost invariant. Suppose that there exists a Borel subset A of G such that $M(B)\ L(\chi_A) \neq L(\chi_A)\ M(B)$. Then $\nu(A \cap r^{-1}(B) \Delta A \cap d^{-1}(B)) > 0$: B is not almost invariant. Conversely, if B is not almost invariant, then either $r^{-1}(B)\setminus d^{-1}(B)$ or $d^{-1}(B)\setminus r^{-1}(B)$ has positive ν-measure and contains a non-singular Borel G-set S with $\mu(r(S)) > 0$. Then $M(\chi_B)V(s) = V(s)$ and $V(s)M(\chi_B) = 0$. Since $V(s)$ is the weak closure of $\{L(f) : f \in C_c(G)\}$, there exists $f \in C_c(G)$ such that $M(\chi_B)L(f) \neq L(f)M(\chi_B)$. Since we assume that L is a factor representation, this shows that μ is ergodic. From c, we conclude that the Hilbert bundle $\mathcal{K}$ is homogeneous, hence isomorphic to a constant Hilbert bundle, so that we can write $H = L^2(G^0,\mu,K)$.

e. We show that L satisfies the inequality $|(L(f)\xi,\eta)| \leq \int|f|d\nu_0 \quad \|\xi\| \quad \|\eta\|$ for every $\xi,\eta \in K$ where $\zeta \in K$ is identified with the constant section $u \to \zeta$. Let ξ and η be fixed unit vectors in K. Since by a. the measure $f \to (L(f)\xi,\eta)$ is absolutely continuous with respect to ν_0, there exists a Borel function c such that $(L(f)\xi,\eta) = \int f(x)\ c(x)d\nu_0(x)$ for every $f \in B(G)$. We have to prove that $|c(x)| \leq 1$ ν a.e. If not, there exist $a > 1$ such that $\nu\{x \in G : |c(x)| > a\} > 0$ and we may find a Borel set A contained in $\{x \in G : |c(x)| > a\}$of positive ν-measure and such that for $u \in r(A)$ and every open set V which meets A^u, $\lambda^u(V \cap A) > 0$. Proceeding as in a, we find a non-singular Borel G-set S of positive μ-measure contained in A and a sequence (U_n,e_n) where U_n is a Borel set contained in A and e_n a non-negative function in B(G) vanishing off U_n such that

(i) $\int e_n d\lambda^u = 1$ for $u \in r(A)$;

(ii) U_n shrinks to S when $n \to \infty$; and

(iii) for every $y \in U_n$, $D(s^{-1}y) \leq b^2$ where $1 < b < a$.

It is possible to fulfill this last condition because any neighborhood of a subset of G^0 of positive μ-measure contains a subset of positive ν-measure where $D \leq b^2$. Let $f_n(y) = \delta^{1/2}(r(y),s)\ e_n(y)|\bar{c}(y)/|c(y)||$. Then

$$\begin{aligned}(L(f_n)\xi,\eta) &= \int f_n(y)\cdot c(y)\ D^{-1/2}(y)d\nu(y)\\ &= \int\delta^{1/2}(r(y),s)\ D^{-1/2}(r(y)s)\ e_n(y)|c(y)|D^{-1/2}(s^{-1}y)\ d\nu(y)\\ &= \iint\delta^{1/2}(u,s)\ D^{-1/2}(us)\ \ e_n(y)|c(y)|\ D^{-1/2}\ (s^{-1}y)\ d\lambda^u(y)\ d\mu(u)\\ &= \iint\Delta^{1/2}(u,s)\ \ e_n(y)|c(y)|\ D^{-1/2}\ (s^{-1}y)\ d\lambda^u(y)\ d\mu(u)\end{aligned}$$

by (1.3.20) and this dominates

$$ab^{-1} \int_{r(S)} \Delta^{1/2}(u,s)\, d\mu(u).$$

On the other hand, we know from b that

$$V(s)\xi(u) = \Delta^{1/2}(u,s)\ c(u,s)\ \xi(u\cdot s) \text{ for } u \in r(S)$$

with $c(u,s)$ isometry of $\mathcal{K}_{u\cdot s}$ into $\mathcal{K}_u$. So

$$(V(s)\xi,\eta) = \int \Delta^{1/2}(u,s)(c(u,s)\xi,\eta)\ d\mu(u) \ ; \text{ and}$$

$$|(V(s)\xi,\eta)| \leq \int_{r(S)} \Delta^{1/2}(u,s)\ d\mu(u).$$

This is a contradiction because $L(f_n) = M(h_n)\ V(s)$, where

$$h_n(u) = \int e_n(y)\ \overline{c(y)}/\ |c(y)|\ \ d\lambda^u(y) \text{ satisfies } |h_n|_\infty \leq 1,$$

tends to zero in the weak operator topology. Indeed,

$$f_\alpha * f(x) - h_\alpha sf(x) = \int e_\alpha(y) \frac{\overline{c(y)}}{|c(y)|}\ \ \delta^{1/2}(r(x),s)[\sigma(y,y^{-1}x)f(y^{-1}x) - \sigma(s,s^{-1}x)f(s^{-1}x)]\ d\lambda^{r(x)}(y)$$

tends to zero for every $x \in G$ and every $f \in C_c(G)$, and $|f_\alpha * f(x)| \leq |(sf)(x)|$ and $|h_\alpha sf(x)| \leq |(sf)(x)|$.

f. The conclusion is given by the following lemma, due to P.Hahn ([39], theorem 5.4, page 106).

<u>Lemma</u> (P. Hahn) : Let G and σ be as above. Let L be a representation of $C_c(G,\sigma)$ on $L^2(G^0,\mu,K)$ where μ is a quasi-invariant probability measure and K a separable Hilbert space, such that

(i) $|(L(f)\xi,\eta)| \leq \int |f|d\nu_0\ \ \|\xi\|\ \ \|\eta\|$ for every $\xi,\eta \in K$

(ζ also denotes the constant function $\zeta \in L^2(G^0,\mu,K)$).

(ii) $L(hf) = M(h)L(f)$ for every $h \in C_c(G^0)$ and every $f \in C_c(G)$, where M is multiplication.

Then, L is the integrated form of a σ-representation of G on the constant Hilbert bundle with fiber K over (G^0,μ).

<u>Proof :</u>

a. There exists a weakly measurable map $x \to A(x)$ of G into the bounded operators on K of norm ≤ 1 such that

$$(L(f)\xi,\eta) = \int f(x) \ (A(x)\xi,\eta) \ d\nu_0(x) \text{ for every } \xi,\eta \in K.$$

For the condition (i) means that $f \mapsto (L(f)\xi,\eta)$ is a bounded linear functional on $L^1(G,\nu_0)$ of norm $\leq \|\xi\| \ \|\eta\|$. This gives a map $(\xi,\eta) \mapsto k(\xi,\eta)$ of $K \times K$ into $L^\infty(G,\nu)$. This map is sesquilinear and satisfies $|k(\xi,\eta)|_\infty \leq \|\xi\| \ \|\eta\|$. Using a lifting of $L^\infty(G,\nu)$ into the space $\mathcal{L}^\infty(G,\nu)$ of bounded measurable functions, we obtain for each $x \in G$ a bounded sesquilinear functional on $K \times K$ of norm ≤ 1, hence an operator $A(x)$ of norm ≤ 1. The map $x \to A(x)$ has the required properties.

b. For every $\xi,\eta \in L^2(G^0,\mu,K)$ and for every $f \in C_c(G)$,

$$(L(f)\xi,\eta) = \int f(x) \ (A(x)\xi \circ d(x), \ \eta\circ r(x)) \ d\nu_0(x).$$

Since both sides are bounded sesquilinear functionals on $L^2(G^0 ,\mu, K)$ (cf. 1.7) it suffices to check the equality on the algebraic tensor product $L^2(G^0,\mu) \otimes K$ and by sesquilinearity on elements of the form $h(u)\xi$, where $h \in C_c(G^0)$ and $\xi \in K$:

$$\begin{aligned}(L(f)h\xi,k\eta) &= (L(f)M(h)\xi, \ M(k)\eta) \\ &= (L(k^* fh)\xi,\eta) \text{ by (ii)} \\ &= \int \overline{k} \circ r(x) \ f(x) \ h\circ d(x) \ (A(x)\xi,\eta) \ d\nu_0(x) \\ &= \int f(x) \ (A(x) \ h\circ d(x)\xi, \ k\circ r(x)\eta) \ d\nu_0(x)\end{aligned}$$

c. A satisfies $A(x)^* = \overline{\sigma}(x,x^{-1}) \ A(x^{-1})$ for ν a.e. x. For all $\xi,\eta \in K$ and $f \in C_c(G)$

$$\begin{aligned}(L(f^*)\xi,\eta) &= \int f^*(x) \ (A(x)\xi,\eta) \ d\nu_0(x) \\ &= \int \overline{f}(x^{-1}) \ \overline{\sigma}(x,x^{-1}) \ (A(x^{-1})(\xi,\eta) \ d\nu_0(x) \\ &= \int \overline{f}(x) \ \overline{\sigma}(x,x^{-1})(A(x^{-1})(\xi,\eta) \ d\nu_0(x) \text{ (because } \nu_0 \text{ is symmetric), and}\end{aligned}$$

$$(\xi,L(f)\eta) = \int \overline{f}(x) \ (\xi, \ A(x)\eta) \ d\nu_0(x)$$

Hence $(A(x)^* \xi,\eta) = \overline{\sigma}(x,x^{-1}) \ (A(x)\xi,\eta)$ for ν a.e.x. Since K is separable, we obtain the result.

d. The function A satisfies $A(x)A(y) = \sigma(x,y) \ A(xy) \ \ \nu^2$ a.e. (x,y).

For all $\xi,\eta \in L^2(G^0,\mu,K)$ and $f,g \in C_c(G)$

$$L(g)\xi(u) = \int g(y) \ A(y) \ \xi\circ d(y) \ D^{-1/2}(y) \ d\lambda^u(y) \text{ for } \mu \text{ a.e. u by b and}$$

$$(L(f)L(g)\xi,\eta) = \int f(x)g(y) \ (A(x)A(y)\xi\circ d(y),\eta\circ r(x)) \ D^{-1/2}(xy) \ D^{-1}(y) \ d\nu^2(xy).$$

(These computations have already been done in 1.7.) On the other hand,

$$(L(f * g)\xi,\eta) = \int f(xy)g(y^{-1})\sigma(xy,y^{-1}) \, (A(x)\xi \circ d(x), \eta\circ r(x)) \, D^{1/2}(x) \, d\nu^2(x,y).$$

After the change of variable $(x,y) \to (xy,y^{-1})$, this is equal to

$$\int f(x) \, g(y) \, \sigma(xy,y^{-1}) \quad (A(xy)\xi\circ d(y), \eta\circ r(x)) \, D^{1/2}(xy) \, D^{-1}(y) \, d\nu^2(x,y)$$

Using the density of $C_c(G) \otimes C_c(G)$ in $C_c(G^2)$, we obtain for all $\xi,\eta \in K$

$$(A(x)A(y)\xi, \eta) = \sigma(x,y) \, (A(xy)\xi,\eta) \quad \text{for } \nu^2 \text{ a.e. } (x,y)$$

since K is separable, we obtain our result.

e. Let S be a non-singular Borel G-set. Since A cannot be evaluated on S, we consider instead a function B defined as follows. It results from d that $A(xy)A(y^{-1}) = \sigma(xy,y^{-1}) \, A(x)$ for ν^2 a.e. (x,y) or equivalently, for ν a.e. x and $\lambda^{d(x)}$ a.e. y. Let $B(x) = \int A(xy) \, A(y^{-1}) \, \overline{\sigma}(xy,y^{-1}) \, f(d(x),y) \, d\lambda^{d(x)}(y)$, where f is a positive measurable function on $G^0 \times G$ such that

$$\int f(u,y) \, d\lambda^u(y) = 1 \text{ for every } u \in G^0.$$

Then, $B(x) = A(x)$ for ν a.e. x. Let us show that $B(xs^{-1})B(s) = \sigma(xs^{-1},s) \, B(x)$ for ν a.e.x. (As usual s in $B(s)$ and in $\sigma(xs^{-1},s)$ stands for $d(xs^{-1})s$). By quasi-invariance of ν under S, it results from d that for ν a.e. $x \in d^{-1}(d(s))$

$A(xs^{-1})A(y) = \sigma(xs^{-1},y)A(xs^{-1}y)$ for $\lambda^{d(xs^{-1})}$ a.e. y and

$B(xs^{-1}) = A(xs^{-1})$. Therefore

$$\begin{aligned}
B(xs^{-1})B(s) &= A(xs^{-1}) \int A(sy)A(y^{-1}) \, \overline{\sigma}(sy,y^{-1})f(d(x),y)d\lambda^{d(x)}(y) \\
&= A(xs^{-1}) \int A(y)A(y^{-1}s) \quad \overline{\sigma}(y,y^{-1}s)f(d(x),s^{-1}y) \, d\lambda^{d(x)\cdot s^{-1}}(y) \\
&= \int A(xs^{-1}y)A(y^{-1}s) \, \sigma(xs^{-1},y) \, \overline{\sigma}(y,y^{-1}s)f(d(x),s^{-1}y) \, d\lambda^{d(x)\cdot s^{-1}}(y) \\
&= \int A(xy)A(y^{-1})\sigma(xs^{-1},sy) \, \overline{\sigma}(sy,y^{-1})f(d(x),y) \, d\lambda^{d(x)}(y) \\
&= \sigma(xs^{-1},s)B(x).
\end{aligned}$$

f. $A(x)$ is a unitary operator for ν a.e. x. Hahn's proof uses the von Neumann selection lemma. We will use instead the existence of sufficiently many non-singular G-sets. The set $E = \{x \in G : B(x) \text{ is not unitary}\}$ is a measurable set. Suppose that it has positive ν-measure. Then, it contains a non-singular Borel G-set S of positive μ -measure. Let us define $V^1(s)$ on $L^2(G^0,\mu,K)$ by

$$(*) \quad V^1(s) \, (u) = \Delta^{1/2}(u,s)B(us)\xi(u\cdot s) \text{ if } u \in r(S) \text{ and}$$
$$= 0 \qquad \text{otherwise.}$$

Then for every $f \in C_c(G)$,

$$L(f)V^1(s)\xi(u) = \int f(x)B(x)\,\Delta^{1/2}(d(x),s)\,B(dx)s)\,\xi(d(x)\cdot s)D^{-1/2}(x)\,d\lambda^u(x)$$

for μ a.e. u.

We change the variable x into xs^{-1} to obtain

$$\int f(xs^{-1})B(xs^{-1})\,\Delta^{1/2}(d(x)\cdot s^{-1},s)B(s)\xi(d(x))\,D^{-1/2}(xs^{-1})\,\delta(d(x),s^{-1})d\lambda^u(x).$$

We use $\Delta(u,s^{-1})D(us^{-1}) = \delta(u,s^{-1})$ for μ a.e. u (1.3.20) to obtain

$$\int f(xs^{-1})\,\delta^{1/2}(d(x),s^{-1})\,B(xs^{-1})\,B(s)\,\xi(d(x))\,D^{-1/2}(x)\,d\lambda^u(x).$$

Finally by d, this yields

$$\int f(xs^{-1})\,\delta^{1/2}(d(x),s^{-1})\,\sigma(xs^{-1},s)\,B(x)\quad \xi\circ d(x)\,D^{-1/2}(x)\,d\lambda^u(x)$$

$$= \int fs(x)\,A(x)\quad \xi\circ d(x)\,D^{-1/2}(x)\,d\lambda^u(x)$$

$$= L(fs)\xi(u).$$

Hence $L(f)V^1(s) = L(fs) = L(f)V(s)$ for every f, so that $V^1(s) = V(s)$.In particular $V^1(s)$ is a non-zero partial isometry with range $M(\chi_{r(S)})$. Comparing (*) and b of the proof of the theorem, we see that B(us) is a unitary operator for μ a.e. u in r(S). This is a contradiction. Hence, for ν a.e. x, B(x) and consequently A(x) are unitary. It suffices to modify A on a null set so that it becomes unitary for every x to fulfill all the conditions of definition 1.6.

Q.E.D.

<u>1.22. Corollary :</u> Under the assumptions of the theorem, every representation of $C_c(G,\sigma)$ on a separable Hilbert space is bounded.

<u>1.23. Corollary :</u> Under the assumption of the theorem, the integrating process 1.7 establishes a bijective correspondence between (σ,G)- Hilbert bundles and separable Hermitian $C^*(G,\sigma)$-modules which preserves intertwining operators.

<u>Proof :</u> We have to prove that two σ-representations $(\mu,\mathcal{H},L)$ and $(\mu',\mathcal{H}',L')$ which give unitarily equivalent integrated representations are equivalent. Let L and L' be representations of $C_c(G,\sigma)$ on $\Gamma(\mathcal{H})$ and $\Gamma(\mathcal{H}')$ whose restrictions to $C_c(G^0)$ are multiplications M and M'. If ϕ is an isometry of $\Gamma(\mathcal{H})$ onto $\Gamma(\mathcal{H}')$ which intertwines L and L', it also intertwines M and M'. Therefore, the scalar spectral measures μ and μ' of M and M' are equivalent and there exists a measurable field $u \mapsto \phi(u)$ where $\phi(u)$ is an

isometry of $\mathcal{K}_u$ onto $\mathcal{K}'_u$, decomposing ϕ. The relation $\phi L(f) = L'(f)\phi$ becomes $\int f(x)(\phi \circ r(x)L(x)\xi \circ d(x), \eta \circ r(x))d\nu_0(x) = \int f(x)(L'(x)\phi \circ d(x)\xi \circ d(x), \eta \circ r(x))\, d\nu_0(x)$ where we have assumed $\mu = \mu'$. This gives $\phi \circ r(x)\, L(x) = L'(x)\phi \circ d(x)$ for ν a.e. x.

Q.E.D.

The σ-representations of a group G can be considered as ordinary representations of the extension group G^σ. This leads to an alternate definition of $C^*(G,\sigma)$, for a groupoid G. Let G^σ denote the extension $\mathbb{T} \times G$ of $\mathbb{T}$ by G defined by the 2-cocycle $\sigma \in Z^2(G, \mathbb{T})$. Recall (1.1.12) that

$$(s,x)\ (t,y) = (st\sigma(x,y), xy)$$
$$(s,x)^{-1} = (s^{-1}\sigma(x,x^{-1})^{-1}, x^{-1})$$

Its unit space can be identified with G^0. It is a locally compact groupoid with the product topology and it has the left Haar system $\{h \times \lambda^u\}$ where h is the Haar measure of $\mathbb{T}$.

<u>1.22. Proposition :</u> The C^*-algebra $C^*(G,\sigma)$ is the quotient of $C^*(G^\sigma)$ by the kernel I_σ of the representation L of $C^*(G^\sigma)$, obtained by integration, which satisfy $L(tf) = tL(f)$ for any $t \in \mathbb{T}$, $f \in C_c(G^\sigma)$ and where $tf(s,x) = f(t^{-1}s,x)$.

<u>Proof :</u> The map π from $C_c(G^\sigma)$ to $C_c(G,\sigma)$ given by the formula $\pi f(x) = \int f(s,x)\, ds$ is a $*$-homomorphism. Indeed,

$$\pi(f*g)(x) = \int f*g(s,x)ds$$
$$= \iiint f(st\sigma(x,y),xy)g(t^{-1}\sigma(y,y^{-1})^{-1},y^{-1})dt d\lambda^{d(x)}(y)ds.$$

One makes the changes of variable $u = st\sigma(x,y)$ and $v = t^{-1}\sigma(y,y^{-1})^{-1}$ to obtain

$$\iiint f(u,xy)\ g(v,y^{-1})\ uv\sigma(y,y^{-1})\sigma(x,y)^{-1}dudvd\lambda^{d(x)}(y)$$
$$= \int (\int f(u,xy)udu)\ (\int g(v,y^{-1})vdv)\sigma(xy,y^{-1})\ d\lambda^{d(x)}(y)$$
$$= \pi(f) * \pi(g)\ (x).$$

Moreover,

$$\pi(f^*)(x) = \int f^*(s,x)sds$$
$$= \int \bar{f}(s^{-1}\sigma(x,x^{-1})^{-1},x^{-1})sds$$
$$= \int \overline{f(t,x^{-1})\ \bar{t}\ dt\ \bar{\sigma}(x,x^{-1})}$$
$$= \pi(f)(x^{-1})\ \sigma(x,x^{-1})$$

$= \pi(f)^* (x)$.

Since π is bounded with respect to the L_I norms, i.e.
$\|\pi(f)\|_I \leq \|f\|_I$, because

$\int |\pi(f)(x)| d\lambda^u(x) \leq \iint |f(x,s)| \, ds d\lambda^u(x)$, it follows that it is bounded with respect to the C^*-norms and extends to a $*$-homomorphism from $C^*(G^\sigma)$ to $C^*(G,\sigma)$. It is onto since its image is closed and contains $C_c(G,\sigma)$. If L is a representation of $C^*(G,\sigma)$, $L\circ\pi$ is a representation of $C^*(G^\sigma)$ which satisfies

$L\circ\pi\ (tf) = L(t\pi(f)) = t\ L\circ\pi(f)$ since $\pi(tf) = t\pi(f)$.

Conversely, if L is a representation of $C^*(G^\sigma)$, satisfying this relation and which is of the form

(*) $(L(f)\xi,\eta) = \iint f(s,x)\ (L(s,x)\xi\circ d(x),\ \eta\circ r(x))\ d\nu_0(x)ds$,

then $L(s,x)$ satisfies

$L(ts,x) = tL(s,x)$

for $h \times h$ a.e. (t,s) and ν a.e. x, as one sees from the equation $L(gf) = \tau(g)L(f)$ where $g \in C_c(\mathbb{T})$,

$(gf)(s,x) = \int g(t)f(t^{-1}s,x)dt$ and $\tau(g) = \int g(t)tdt$.

$L(s,x)$ can be replaced by

$L'(s,x) = \int L(ts,x)\ \bar{t}\ dt$

without changing (*). Then it satisfies

$L(ts,x) = tL(s,x)$ for every (t,s) and ν a.e. x. Thus

$(L(f)\xi,\eta) = \iint f(s,x)sds\ (L(e,x)\xi\circ d(x),\eta\circ r(x))\ d\nu_0(x)$,

so that L factors through π.

Q.E.D.

2. Induced Representations

Let G be a locally compact groupoid with Haar system $\{\lambda^u\}$ and H a closed subgroupoid G containing G^0 and admitting a Haar system $\{\lambda_H^u\}$, and $\sigma \in Z^2(G,\mathbb{T})$ a continuous 2-cocycle. Just as in the case of groups, a σ-representation of H may be induced to a

σ-representation of G. We describe this process below. Rieffel's version [54] of induced representations is particularly well adapted to this context and this section imitates the exposition he gives in the case of groups. We first need some topological results which are well known in the group situation.

2.1. Proposition : Let G and H be as above and consider the relation on G defined by $x \sim y$ iff $d(x) = d(y)$ and $xy^{-1} \in H$.

(i) It is an equivalence relation.

(ii) The quotient topology on the quotient space $H\backslash G$ is Hausdorff.

(iii) The quotient map $r : G \to H\backslash G$ is open.

(iv) The quotient space $H\backslash G$ is locally compact.

(v) The domain map d induces a continuous and open map from $H\backslash G$ onto G^0.

Proof :

(i) Clear.

(ii) Since H is closed, the set $\{(x,y) \in G^2 : xy \in H\}$ is closed in G^2, hence in $G \times G$. The graph of the relation is the image by θ of this set, where θ is the homeomorphism $(x,y) \mapsto (x,y^{-1}) : G \times G \to G \times G$

(iii) Let O be an open set G ; we have to show that its saturation HO is also open. Let hx be a point in HO with $h \in H$ and $x \in O$. There exists a non-negative function $\phi \in C_c(H)$ such that $\phi(h) \neq 0$ and a non-negative function $g \in C_c(G)$ such that $g(x) \neq 0$ and $\operatorname{supp} g \subset O$. The same argument as in 1.1 shows that the function $\phi\cdot g$ defined by

$$\phi\cdot g\ (y) = \int \phi(k)\, g(k^{-1}y)\, d\lambda_H^{d(y)}\ (k)$$

is continuous on G ; therefore $\{y : \phi\cdot g(y) \neq 0\}$ is an open set ; since it contains hx and is contained in HO, we are done.

(iv) This results from (ii) and (iii).

(v) This is clear since $d : G \to G^0$ is continuous and open.

Q.E.D.

2.2. Lemma : There exists a Bruhat approximate cross-section for G over $H\backslash G$, that is, a non-negative continuous function b on G whose support has compact intersection with the saturation HK of any compact subset K of G and is such that for every $x \in G$, $\int b(h^{-1}x)\, d\lambda_H^{r(x)}(h) = 1$.

Proof : By [6], Lemme 1, page 96, there exists a non-negative continuous function g non-zero on every equivalence class and whose support has compact intersection with the saturation of any compact subset of G. The function $g°$ defined by $g°(x) = \int g(h^{-1}x)\, d\lambda_H^{r(x)}(h)$ is continuous and strictly positive. The function $b = g/g°$ is a Bruhat approximate cross section for G over $H\backslash G$.

Q.E.D.

2.3. Proposition : Let G and H be as above and consider the relation $\sim$ on G^2 defined by $(x,y) \sim (x',y')$ iff $y = y'$ and $xx'^{-1} \in H$.

(i) It is an open Hausdorff equivalence relation and the quotient space $H\backslash G^2$ with quotient topology is locally compact.

(ii) The relation is compatible with the groupoid structure of G^2, so that $H\backslash G^2$ is a locally compact groupoid, its unit space may be identified with $H\backslash G$.

(iii) The groupoid $H\backslash G^2$ has a Haar system, namely $\{\delta_{\dot{x}} \times \lambda^{d(\dot{x})},\ \dot{x} \in H\backslash G\}$.

Proof :

(i) This is verified as 2.1, in fact $H\backslash G^2 = \{(\dot{x},y) \in H\backslash G \times G : d(\dot{x}) = r(y)\}$.

(ii) The composable pairs in G^2 are $((x,y),(xy,z))$. Therefore if $(x',y) \sim (x,y)$, then $(x'y,z) \sim (xy,z)$ and $(x',y)\,(x'y,z) = (x',yz) \sim (x,yz) = (x,y)(xy,z)$. Hence we may define the following groupoid structure on $H\backslash G^2$. The composable pairs are $(\dot{x},y)$, $(\dot{\overline{xy}},z)$, $(\dot{x},y)\,(\dot{\overline{xy}},z) = (\dot{x},yz)$ and the inverse of $(\dot{x},y)$ is $(\dot{\overline{xy}},y^{-1})$. Finally, by definition of the quotient topology, the multiplication and inverse maps are continuous. Since $(\dot{x},y)\,(\dot{x},y)^{-1} = (\dot{x},yy^{-1}) = (\dot{x},d(\dot{x}))$, we may identify the unit space of $H\backslash G^2$ and $H\backslash G$.

(iii) This is clear ; here $\int f\, \delta_{\dot{x}} \times d\lambda^{d(\dot{x})} = \int f(\dot{x},y)\,d\lambda^{d(\dot{x})}(y)$

Q.E.D.

Notation : Let σ be a continuous 2-cocycle in $Z^2(G,\mathbb{T})$, one can associate with it continuous 2-cocycles on H and on $H\backslash G^2$ respectively in the following way. On H, σ denotes its restriction to H. On $H\backslash G^2$, σ is defined by $\sigma(\dot{x},y,z) = \sigma(y,z)$ (we write $(\dot{x},y,z)$ instead of $((\dot{x},y),\,(\dot{\overline{xy}},z))$). The cocycle property is easily checked.

For $\phi \in C_c(H,\sigma)$ and $f \in C_c(G)$, let us define

$$\phi \cdot f\ (x) = \int \phi(h)\ f(h^{-1}x)\ \sigma(h,h^{-1}x)\ d\lambda_H^{r(x)}(h), \text{ and}$$

$$f \cdot \phi\ (x) = \int f(xh)\ \phi(h^{-1})\ \sigma(xh,h^{-1})\ d\lambda_H^{d(x)}(h\ .$$

For $\psi \in C_c(H\backslash G^2,\sigma)$ and $f \in C_c(G)$, let us define

$$\psi \cdot f\ (x) = \int \psi(\dot{x}^{-1},xy)\ \ f(y^{-1})\ \sigma(xy,y^{-1})d\lambda^{d(x)}(y), \text{ and}$$

$$f \cdot \psi\ (x) = \int f(y)\ \psi(\dot{y},y^{-1}x)\ \sigma(y,y^{-1}x)\ d\lambda^{r(x)}(y)\ .$$

2.4. Proposition :

(i) The space $C_c(G)$ is a $C_c(H,\sigma)$-bimodule and a $C_c(H\backslash G^2,\sigma)$-bimodule ; and the actions of $C_c(H,\sigma)$ and $C_c(H\backslash G^2,\sigma)$ on opposite side commute.

(ii) The algebra $C_c(H,\sigma)$ acts as a $*$-algebra of double centralizers on the algebra $C_c(G,\sigma)$, this action extends to the C^*-algebra $C^*(G,\sigma)$ and gives a $*$-homomorphism of $C^*(H,\sigma)$ into the multiplier algebra of $C^*(G,\sigma)$.

Proof :

(i) One has first to check that, with above notations, ϕf, $f\phi$, $\psi\cdot f$ and $f\cdot\psi$ are indeed in $C_c(G)$. This is done in exactly the same fashion as in proposition 1.1. The verification of the various associativity relations,

$$(\phi * \psi)\ \cdot f = \phi\cdot(\psi\ \ \cdot f) \text{ for } f \in C_c(G)$$

and ϕ,ψ both in $C_c(H,\sigma)$ or in $C_c(H\backslash G^2,\sigma)$, the analogous relation for the action on the right, and

$$\phi\ \cdot(f \cdot \psi) = (\phi \cdot f) \cdot \psi \text{ for } f \in C_c(G)$$

and ϕ,ψ in $C\ (H,\sigma)$ or in $C\ (H\backslash G^2,\sigma)$, is straightforward but tedious. Let us check one of them as an example. Suppose $f \in C_c(G)$ and $\phi,\psi \in C_c(H\backslash G^2,\sigma)$. Then

$$\phi * \psi\ (\dot{x},y) = \int \phi(\dot{x},yz)\psi(\dot{\overline{xyz}},z^{-1})\ \sigma(yz,z^{-1})d\lambda^{d(y)}(z), \text{ and}$$

$$\begin{aligned}
f\cdot(\phi * \psi)\ (x) &= \int f(y)\phi\quad \psi(\dot{y},y^{-1}x)\ \sigma(y,y^{-1}x)d\lambda^{r(x)}(y)\\
&= \iint f(y)\ \phi(\dot{y},y^{-1}xz)\ \psi(\dot{\overline{xz}},z^{-1})\ \sigma(y,y^{-1}x)\ \sigma(y^{-1}xz,z^{-1})\\
&\quad \cdot d\lambda^{d(x)}(z)d\lambda^{r(x)}(y)\\
&= \iint f(y)\ \phi(\dot{y},y^{-1}xz)\ \psi(\dot{\overline{xz}},z^{-1})\ \sigma(y,y^{-1}xz)\ \sigma(xz,z^{-1})\\
&\quad \cdot d\lambda^{r(x)}(y)\ d\lambda^{d(x)}(z)\\
&= \int f \cdot \phi(xz)\ \psi(\dot{\overline{xz}},z^{-1})\ \sigma(xz,z^{-1})\ d\lambda^{d(x)}(z)\\
&= \int f \cdot \phi(z)\ \psi(\dot{z},z^{-1}x)\ \sigma(z,z^{-1}x)d\lambda^{r(x)}(z)
\end{aligned}$$

$= (f \cdot \phi) \cdot \psi\ (x)$.

(ii) We have to check the equations

$f * (\phi \cdot g) = f \cdot \phi * g$ and $(\phi \cdot f)^* = f^* \cdot \phi^*$

for f, $g \in C_c(G)$ and $\phi \in C_c(H)$. This is done as above. To prove that this action extends extends to $C^*(G,\sigma)$, one can introduce the Banach algebra $L_{I,r}(G,\sigma)$, the completion of $C_c(G,\sigma)$ for the norm $\| \|_{I,r}$. It has a bounded left approximate identity. Thus, if L is a bounded representation of $C_c(G,\sigma)$, there is a unique bounded representation L_H, called the restriction of L to $C^*(H,\sigma)$, such that $L(\phi \cdot f) = L_H(\phi)L(f)$ and $L(f \cdot \phi) = L(f)\, L_H(\phi)$. What makes the proof go is the inequality $\|\phi \cdot f\|_{I,r} \leq \|\phi\|_{I,r}\|f\|_{I,r}$ which is obtained as in 1.6. This gives a faithful $*$-homomorphism of $C_c(H,\sigma)$ into the multiplier algebra of $C^*(G,\sigma)$ which is norm-decreasing when $C_c(H,\sigma)$ has the $\| \|_I$ norm. Hence it extends to a $*$-homomorphism of $C^*(H,\sigma)$ into the multiplier algebra of $C^*(G,\sigma)$.

Q.E.D.

Let $X = C_c(G)$, $B = C_c(H,\sigma)$ and $E = C_c(H\backslash G^2,\sigma)$; view X as a left E- and right B-bimodule. One would like to exhibit X as an E-B imprimitivity bimodule (definition 6.10 of [63]). I did not succeed in doing that except in particular cases. The candidates for E and B-valued inner products on X are

$$<f , g>_B (h) = \int \bar{f}(x^{-1})\, g(x^{-1}h)\bar{\sigma}(x,x^{-1})\, \sigma(x,x^{-1}h)d\lambda^{r(h)}(x) \text{ and}$$

$$<f , g>_E(\dot{x},x^{-1}y) = \int f(x^{-1}h)\, \bar{g}(y,h)\, \bar{\sigma}(y^{-1}h)\, \bar{\sigma}(y^{-1}h,h^{-1}y)\, \sigma(x^{-1}h,h^{-1}y)d\lambda_H^{r(x)}(h).$$

(By left invariance of the Haar system, the right hand side depends on x only). The algebraic relations

$$<f , gb>_B = <f , g>_B b \qquad (f , g \in X,\ b \in B)$$

$$<ef,g>_E = e<f , g>_E \qquad (e \in E)$$

$$<f,gb>_E = <fb^* ,g>_E$$

$$<ef,g>_B = <f,e^* g>_B$$

$$f<g,f'>_B = <f,g>_E f' \qquad (f,\ g,\ f' \in X)$$

are satisfied, as one may check in the same fashion as above.

2.5. Lemma :

(i) The linear span of the range of $<,>_E$ contains an approximate left identity for $C_c(H\backslash G^2,\sigma)$ with the inductive limit topology.

(ii) A similar statement holds for $<,>_B$ and $C_c(H,\sigma)$.

Proof : It is the same proof as in proposition 7.11 of [63] (and in lemma 2, page 201, of [39]).

(i) Let C be a compact subset of $H\backslash G$ and ε a positive number. Choose a compact set K in G such that $\pi(K) = C$. There exists a d-relatively compact (see 1.9) neighborhood N of G^0 in G such that $|\sigma(x^{-1},y) - 1| \leq \varepsilon$ for $y \in N$, $x \in K$ and $r(y) = r(x)$, because σ is continuous and takes the value 1 whenever one of its arguments is a unit. There is a locally finite cover of G consisting of open relatively compact sets (V_i) such that $V_i^{-1}V_i \subset N$ and a partition of unity subordinate to it. Multiply this partition pointwise with a Bruhat approximate cross-section b which has been truncated so that $b \in C_c(G)$ and $\int b(h^{-1}x)d\lambda_H^{r(x)}(h) = 1$ for $x \in K$, and obtain a finite number of non-negative functions of $f_1,\dots,f_n \in C_c(G)$ such that $\text{supp} f_i \subset V_i$ and $\sum_{i=1}^{n} \int f_i(h^{-1}x)d\lambda_H^{r(x)}(h) = 1$ for $x \in K$. For each i, choose $g_i \in C_c(G)$ such that $\text{supp} g_i \subset V_i$, $\int |g_i(y)|d\lambda^u(y) = 1$ for $u \in r\ \{x : f_i(x) \neq 0\}$, and $|g_i(y)| = \overline{g_i}(y)\,\overline{\sigma}(y^{-1},y)$. Let $\theta_{(c,\varepsilon,N)} = \sum_{i=1}^{n} <\tilde{f}_i,\tilde{g}_i>$, where $\tilde{f}(x) = f(x^{-1})$. Since $<\tilde{f}_i,\tilde{g}_i>(\dot{x},y) = \int f_i(h^{-1}x)\,\overline{g_i}(h^{-1}xy)\,\overline{\sigma}(y^{-1}x^{-1}h,h^{-1}xy)\,\sigma(x^{-1}h,h^{-1}xy)d\lambda_H^{r(x)}(h)$, $\theta_{(c,\varepsilon,N)}$ satisfies

(a) $\theta_{(c,\varepsilon,N)}(\dot{x},y) = 0$ if $y \notin N$, and

(b) $|\int \theta_{(c,\varepsilon,N)}(\dot{x},y)d\lambda^{d(x)}(y) - 1| \leq \varepsilon$ if $\dot{x} \in C$.

It results from the proof of proposition 1.9 that the net $\{\theta_{(c,\varepsilon,N)}\}$ directed by $(c,\varepsilon,N) \prec (c',\varepsilon',N')$ iff $c \subset c'$, $\varepsilon \geq \varepsilon'$ and $N \supset N'$ is a left approximate identity for $C_c(H\backslash G^2,\sigma)$.

(ii) This is done in a similar manner. Let K be a compact subset of G^0, ε a positive number and N an r-relatively compact neighborhood of G^0 in G. One can find an r-relatively compact neighborhood U of G^0 and non-negative continuous functions f and g on G such that $UU^{-1} \subset N$, the support of g is compact and contained in U, while $\int g(x)d\lambda^u(x) = 1$ for $u \in K$, the support of f is contained in U and has compact intersection with the saturation HL of any compact subset L of G, while $\int f(h^{-1}x)\,d\lambda_H^{r(x)}(h) = 1$ for $x \in r^{-1}(K) \cap U$, and $|\overline{\sigma}(h^{-1}x,x^{-1}h)\sigma(h^{-1}x,x^{-1}) - 1| \leq \varepsilon$ when $x \in U \cap r^{-1}(K)$, $h^{-1}x \in U$ and $h \in H$. Then, one notes that the function $\phi_{(c,\varepsilon,N)}$, defined by

$$\phi_{(N,\varepsilon,K)}(h) = <\tilde{f},\tilde{g}>_B(h)$$
$$= \int f(hx)\, g(x)\, \sigma(hx,x^{-1}h^{-1})\, \sigma(hx,x^{-1}) d\lambda^{d(h)}(x),$$

satisfies

(a) $\phi_{(N,\varepsilon,K)}(h) = 0$ if $h \notin N$ and

(b) $|\int \phi_{(N,\varepsilon,K)}(h) d\lambda_{Hu}(h) - 1| \leq \varepsilon$ if $u \in K$.

Therefore, the net $\{\phi_{(n,\varepsilon,K)}\}$ is a right approximate identity for $C_c(H,\sigma)$.

Q.E.D.

2.6. Corollary :

(i) The linear span of the range of $<,>_E$ is dense in $C_c(H\backslash G^2,\sigma)$ and in $C^*(H\backslash G^2,\sigma)$

(ii) The linear span of the range of $<,>_B$ is dense in $C_c(H,\sigma)$ and in $C^*(H,\sigma)$.

Remark : It seems difficult to construct approximate identities as those obtained in lemma 2, page 201, of [39]. In the general case, with the notations of the proof, one would need sets V_i's such that $r(V_i) = G^0$. In the case when $H = G^0$, it is not hard to carry the proof through. This is done in the next proposition.

2.7. Proposition : Let $H = G^0$, $B = C_c(G^0)$ and $E = C_c(G^2,\sigma)$. Then $X = C_c(G)$ is an E-B imprimitivity bimodule. In other words, $C^*(G^2,\sigma)$ and $C^*(G^0)$ are strongly Morita equivalent.

Proof : It is not surprising that this result is independent of σ since G^2, being continuously similar to G^0, has trivial cohomology. Thus we may assume that $\sigma \equiv 1$. We have to check the last contitions in the definition of an imprimitivity bimodule. The B-valued inner-product is clearly positive :

$$<f,f>_B(u) = \int |f(y^{-1})|^2 d\lambda^u(y).$$

The E-valued inner-product is positive ; as mentioned before, we can find here an approximate identity for the right action of $C_c(G)$ of the form $<f_K,f_K>_B$; namely, let K be a compact subset of G^0, $g \in C_c(G)$ nonzero on K and $h \in C_c(G^0)$ such that $h(u) = [\int |g(y^{-1})|^2 d\lambda^u]^{-1/2}$ for $u \in K$; then set $f_K = gh$. To complete the proof we need only verify the norm conditions

$$<fb,fb>_E \leq \|b\|^2 <f,f>_E \text{ and } <ef,ef>_B \leq \|e\|^2 <f,f>_B$$

where $e \in E$, $b \in B$ and $f \in X$. But

$$<fb,fb>_E(x,y) = |b\circ r(x)|^2 \ <f,f>_E(x,y) \text{ and}$$

$$\|b\|^2 \ <f,f>_E - <fb,fb>_E = <fc,fc>_E$$

where $c(x) = (\|b\|^2 - |b\circ r(x)|^2)^{1/2}$. Assume that $e \in E$ is non-negative. Then

$$\begin{aligned}<ef,ef>_B(u) &= \int |ef(y^{-1})|^2 d\lambda^u(y)\\ &= \int |\int e(y,y^{-1}z)^{1/2} f(z^{-1}) e(y,y^{-1}z)^{1/2} d\lambda^{r(y)}(z)|^2 d\lambda^u(y)\\ &\leq \iint e(y,y^{-1}z)\ |f(z^{-1})|^2 d\lambda^{r(y)}(z) \int e(y,y^{-1}z) d\lambda^{r(y)}(z)\ d\lambda^u(y)\\ &\leq \sup_y \int e(y,z) d\lambda^{d(y)}(z) \iint e(y,y^{-1}z)|f(z^{-1})|^2 d\lambda^{r(z)}(y)\ d\lambda^u(z)\\ &\leq \|e\|_{I,r} (\sup_z \int e(zy,y^{-1}) d\lambda^{d(z)}(y)) \int |f(z^{-1})|^2 d\lambda^u(z)\\ &\leq \|e\|_{I,r}\ \|e\|_{I,d}\ <f,f>_B(u)\\ &\leq \|e\|_I^2\ <f,f>_B(u).\end{aligned}$$

This gives a $*$-homomorphism $L : E \to L(X)$ where $L(X)$ is the algebra of bounded operators on the pre B-Hilbert space X which is bounded when E has the $\| \ \|_I$-norm ; therefore, it is norm-decreasing. By definition of the C^*-norm on E, $\|L(e)\| \leq \|e\|$ and hence the required inequality.

Q.E.D.

A representation of $C^*(G^0)$ can be induced up to a representation of $C^*(G^2,\sigma)$ by Rieffel's tensor product construction (Corollary 6.15 of [63]) and "restricted" to $C^*(G,\sigma)$, which acts on $C^*(G^2,\sigma)$ as double centralizers (a function on G can be viewed as a function on G^2 depending on the second variable only). Alternatively, the restriction map $P : C_c(G,\sigma) \to C_c(G^0)$ is a generalized conditional expectation ([63], definition 4.12) and so a representation of $C^*(G^0)$ may be induced to $C^*(G,\sigma)$ via P. Let us construct explicitly these representations induced from the unit space (this will be used in 3.2). For simplicity, consider a multiplicity-free representation of $C^*(G^0)$, given by multiplication on the space $L^2(G^0,\mu)$, where μ is a measure on G^0. The space of the induced representation is obtained by completing $C_c(G) \otimes_{C_c(G^0)} C_c(G^0) = C_c(G)$ with respect to the inner product

$$\begin{aligned}<f \otimes h, g \otimes k> &= \int P(g^* * f) h \bar{k}\, d\mu = \int (\int \bar{g}(y^{-1}) f(y^{-1}) d\lambda^u) h(u) \bar{k}(u) d\mu(u)\\ &= \int (f \otimes h) \overline{(g \otimes k)}\, d\nu^{-1}\end{aligned}$$

where, as usual, $\nu = \int d\lambda^u d\mu(u)$. The induced representation, denoted by Indμ acts on $L^2(G,\nu^{-1})$ by convolution on the left : for $f \in C_c(G)$ and $\xi,\eta \in L^2(G,\nu^{-1})$

$$(\mathrm{Ind}\mu(f)\xi,\eta) = \int f(xy)\xi(y^{-1})\overline{\eta}(x)\sigma(xy,y^{-1})\, d\lambda^u(y)d\lambda_u(x)d\mu(u).$$

We have met this representation before, in the case when μ is a quasi-invariant measure (proposition 1.10) : it is the regular representation on μ.

It results from proposition 1.11 that the function defined by $\|f\|_{red}$ = sup $\|L(f)\|$, where L ranges over all representations induced from the unit space, is a C^*-norm on $C_c(G,\sigma)$ dominated by the C^*-norm $\|f\|$.

2.8. Definition : The reduced C^*-algebra $C^*_{red}(G,\sigma)$ of G is the completion of $C_c(G,\sigma)$ for the reduced norm $\|\ \|_{red}$.

It is a quotient of $C^*(G,\sigma)$ since the identity map on $C_c(G,\sigma)$ extends to a $*$-homomorphism of $C^*(G,\sigma)$ onto $C^*_{red}(G,\sigma)$.

Representations induced from more general subgroupoids will only be considered in the context where theorem 1.21 applies. The notion of generalized conditional expectation used the following proposition was introduced by M.Rieffel in [63] (definition 4.12). This is the piece of structure which allows the construction of induced representations.

2.9. Proposition : Assume that G is second contable and that G and H have sufficiently many non-singular Borel G-sets. Then the restriction map from the pre-C^*-algebra $C_c(G,\sigma)$ to the pre-C^*-algebra $C_c(H,\sigma)$ is a generalized conditional expectation.

The following lemma shows the positivity of P ; it is due to Blattner in the case of groups ([3], theorem 1, page 424). We follow here [63], theorem 4.4.

2.10. Lemma : Let $(\mu,\mathcal{H},L)$ be a 1-representation of H ; then, for any $f,g \in C_c(G,1)$ and any $\xi,\eta \in \Gamma(\mathcal{H})$ (the space of square-integrable sections), we have $(L\circ P(g^* * f)\xi,\eta) = \int b(x)(\pi(f,\xi)(x),\pi(g,\eta)(x))d\nu(x)$ where b is a Bruhat approximate cross-section for G over $H\backslash G$, $\nu = \int \lambda^u\, d\mu(u)$ and $\pi(f,\xi)(x) = \quad f(x^{-1}k)L(k)\xi\circ d(k)\, \Delta_H^{-1/2}(k)\, d\lambda_H^{r(z)}(k)$. ($\Delta_H$ denotes the modular function of μ relative to (H,λ_H^u).)

Proof : We have

$$(L\circ P(g^* * f)\xi,\eta) = \int g^* * f(h)\ (L(h)\ \xi\circ d(h),\ \eta\circ r(h))\ d\nu_{HO}(h)$$

where ν_{HO} is the symmetric measure of μ relative to (H, λ_H^u), This, in turn, equals

$$\iint \bar{g}(x^{-1}h^{-1})\ \ f(x^{-1})\ (L(h)\ \xi\circ d(h),\ \eta\circ r(h))\ d\lambda^{d(h)}(x)\ d\nu_{HO}(h)$$

$$= \iint(\int b(k^{-1}x)\ d\lambda_H^{r(x)}\ (k))\ \ \bar{g}(x^{-1}h^{-1}\ f(x^{-1})\ (L(h)\ \xi\circ d(h),\ \eta\circ r(h))\ \ d\lambda^{d(h)}(x)\ d\nu_{HO}(h)$$

$$=\iiint b(k^{-1}x)\ \bar{g}(x^{-1}h^{-1})\ f(x^{-1})\ (L(h)\ \ \xi\circ d(h),\eta\circ r(h))\ d\lambda^{r(k)}(x)\ d\lambda_H^{d(h)}(k)\ d\nu_{HO}(h)$$

The use of Fubini's theorem is legitimate, because the support of the function $(k,x)\to b(k^{-1}x)\ \bar{g}(x^{-1}h^{-1})\ f(x^{-1})$ is compact :
$k^{-1}\ x \in H\ \mathrm{supp}f^* \cap \mathrm{supp}b$. We make the change of variable $x \mapsto kx$ in the last integral to obtain

$$\iiint b(x)\ \bar{g}(x^{-1}k^{-1}h^{-1})\ \ f(x^{-1}k^{-1})\ \ (L(h)\ \xi\circ d(h),\ \eta\circ r(h))d\lambda^{d(h)}(x)\ d\lambda_H^{d(h)}(k)\ d\nu_{HO}(h)\ .$$

We write $d\nu_{HO}(h) = \int \Delta_H^{1/2}\ (h)d\lambda_{Hu}d\mu(u)$ and change the order of integration ; this is justified as above. We get

$$\int b(x)\ \bar{g}(x^{-1}k^{-1}h^{-1})\ f(x^{-1}k^{-1})\ (L(h)\xi\circ d(h),\eta\circ r(h))\ \ \Delta_H^{1/2}(h)\ d\lambda_{Hr(x)}(h)$$
$$d\lambda^{d(k)}(x)\ d\lambda_H^u(k)\ d\mu(u)\ .$$

We make the change of variable $h \to hk^{-1}$, yielding

$$\int b(x)\ \bar{g}(x^{-1}h^{-1})\ \ f(x^{-1}k^{-1})\ (L(hk^{-1})\ \xi\circ r(k),\eta\circ r(h))\ \Delta_H^{1/2}\ (hk^{-1})$$
$$d\lambda_{Hd(k)}(h)\ \ d\lambda^{d(k)}(x)\ d\nu_H(k).$$

We use the fact that $\Delta_H^{-1}(k)\ d\nu_H(k) = d\nu_H^{-1}(k)$ to produce

$$\int b(x)\ \bar{g}(x^{-1}h^{-1})\ f(x^{-1}k^{-1})\ (L(k^{-1})\xi\circ r(k),L(h^{-1})\eta\circ r(h))\ \Delta_H^{1/2}(hk)$$
$$d\lambda_{Hu}(h)\ d\lambda^u(x)\ d\lambda_{Hu}(k)\ d\mu(u).$$

Finally, we change the order of integration and arrive at

$$\int b(x)\ \ (\int (f(x^{-1}k^{-1})L(k^{-1})\xi\circ r(k)\ \Delta_H^{1/2}(k),\ g(x^{-1}h^{-1})$$
$$L(h^{-1})\eta\circ r(h)\ \Delta_H^{1/2}(h))\ \ d\lambda_{Hu}(h)\ d\lambda_{Hu}(k))d\lambda^u(x)\ d\mu(u)\ .$$

The vector-valued function $k \to f(x^{-1}k^{-1})\ L(k^{-1})\xi\circ r(k)\ \ \Delta_H^{1/2}(k)$ is $\lambda_{Hr(x)}$ integrable because it is measurable and its norm is integrable since $|\xi\circ r(k)|\ \Delta_H^{1/2}(k)$ is locally integrable. Hence

$\pi(f,\xi)(x) = \int f(x^{-1}k^{-1})\ L(k^{-1})\xi\circ r(k)\ \Delta_H^{1/2}(k)\ d\lambda_{Hr(x)}(k)$ makes sense and $(\pi(f,\xi)(x),\ \pi(g,\eta)(x))$ is equal to

$$\int\ (f(x^{-1}k^{-1})L(k^{-1})\xi\circ r(k)\ \Delta_H^{1/2}(k),\ g(x^{-1}h^{-1})L(h^{-1})\eta\circ r(h)\ \Delta_H^{1/2}(h))\ d\lambda_{Hr(x)}(h)d\lambda_{Hr(x)}(k).$$

Q.E.D.

Proof of the proposition.

(i) P is self adjoint, because $P(f^*) = P(f)^*$ for $f \in C_c(G,\sigma)$ by simple calculation.

(ii) P is positive, i.e., $P(f^* * f) \geq 0$ in $C^*(H,\sigma)$ for $f \in C_c(G,\sigma)$. To see this, we first consider the case $\sigma = 1$. The lemma tells us that $L(P(f^* * f)) \geq 0$ for any representation L of $C^*(G,1)$ obtained by integration.Since $C^*(G)$ has a faithful family of such representations by theorem 1.21, $P(f^* * f) \geq 0$. To deal with the case of an arbitrary cocycle σ, we consider instead G^σ and H^σ . Since H^σ is a closed subgroupoid of G^σ and G^σ and H^σ satisfy the hypothesis of the proposition, the restriction map Q from $C_c(G^\sigma,1)$ onto $C_c(H^\sigma,1)$ is positive. Since the diagram

$$\begin{array}{ccc} C_c(G^\sigma,1) & \xrightarrow{\quad Q \quad} & C_c(H^\sigma,1) \\ \Big\downarrow \pi & & \Big\downarrow \pi_H \\ C_c(G,\sigma) & \xrightarrow{\quad P \quad} & C_c(H,\sigma) \end{array} \qquad \begin{array}{l} \text{with } \pi f(x) = \int f(s,x)s\,ds \\ \pi_H g(h) = \int g(s,h)s\,ds \end{array}$$

commutes and since π is onto while π_H is continuous, P is also positive.

(iii) P satisfies the expectation property ; i.e., $P(\phi \cdot f) = \phi * P(f)$ for $f \in C_c(G,\sigma)$ and $\phi \in C_c(H,\sigma)$, as can be seen immediately from the definiton of $\phi \cdot f$.

(iv) P is relatively bounded ; that is, for every $g \in C_c(G)$, the map $f \mapsto P(g^* * f * g)$ is bounded with respect to the C^*-norms). To see this, proceed as in proposition 4.10 of [63]. First, one establishes the inequality

$$P(f^* * g^* * g * f) \leq \|g\|^2 \; P(f^* * f) .$$

It suffices to show the inequality when one evaluates both sides against a positive type measure, that is, a measure ν on $C_c(H)$ satisfying $\nu(f^* * f) \geq 0$ for any $f \in C_c(H)$. Via the GNS construction, a positive type measure on G defines a representation of $C_c(G,\sigma)$, hence of $C_c(G,\sigma)$ by theorem 1.20 ; in particular, it is continuous with respect to the C^*-topology. Therefore, $\nu \circ P$ is a positive type measure on $C_c(G,\sigma)$. The corresponding representation is given by convolution on the left ;

$$\|g * f\|^2_{\nu \circ P} \leq \|g\|^2 \; \|f\|^2_{\nu \circ P} \qquad \text{for } g,f \in C_c(G),$$

where $(\;,\;)_{\nu \circ P}$ is the inner-product defined by $\nu \circ P$:

$$\nu \circ P((g * f)^* * (g * f)) \leq \|g\|^2 \; \nu \circ P\,(f^* * f) .$$

Then, one applies the generalized Cauchy-Schwarz inequality of proposition 2.9 of [63] to conclude.

(v) Let (e_k) be an approximate left identity for $C_c(G,\sigma)$ with the inductive limit topology. Then, for any $f \in C_c(G,\sigma)$, $(f - e_k * f)^* * (f - e_k * f)$ tends to 0 in $C_c(G,\sigma)$ and its restriction to H tends to 0 in $C_c(H,\sigma)$, hence in the C^*-norm.

(vi) The range of P is $C_c(H,\sigma)$.

(vii) P is clearly faithful ; $f^* * f = 0$ on $H \Longrightarrow f^* * f = 0$ on $G^0 \Longrightarrow f = 0$.

Q.E.D.

This proposition allows a partial answer to a question that has been avoided until now. Given a locally compact groupoid, we have assumed the existence of a Haar system and kept it fixed. Most notions introduced, such as quasi-invariance or the convolution product, depend explicitly on the choice of such a Haar system. What is the role of this choice and can we find notions independent of it ?

2.11. Corollary : Let G be a second countable locally compact groupoid, (λ_i^u) $i = 1,2$ two Haar systems with respect to which G has sufficiently many non-singular Borel G-sets and let σ be a continuous 2-cocycle. Then, the corresponding C^*-algebras $C^*(G_1,\sigma)$ and $C^*(G_2,\sigma)$ are strongly Morita equivalent.

Proof : We set $G = G_1$ and view G_2 as the subgroup H. Then $H\backslash G^2 = G_1$. We can use proposition 2.9 to show that $X = C_c(G)$ is indeed an E-B imprimitivity bimodule with $E = C_c(H\backslash G^2,\sigma)$ and $B = C_c(H,\sigma)$ as before. Propositon 2.9 gives the positivity of $\langle f,f\rangle_B$ as well as the norm condition $\langle ef,ef\rangle_B \le \|e\|_1 \langle f,f\rangle_B$ for $e,f \in C_c(G)$. By symmetry, similar statements hold for E.

Q.E.D.

2.12. Example : Let X be a second countable locally compact space. We have defined (1.3.28.c) the transitive groupoid on the space X as $G = X \times X$, with the groupoid structure given in 1.1.2 (ii) and the product topology. We know that a Haar system on G is defined by a measure α of support X. If X is uncountable and α is non-atomic, then G has sufficiently many non-singular Borel G-sets. Let us fix α. Since the class of α is the only invariant measure class and any representation of G is a multiple of the one-dimensional trivial representation, the corresponding C^*-algebra is isomorphic to the algebra of compact operators on a separable Hilbert space. Thus two measures α_1 and α_2 give isomorphic C^*-algebras but there is no canonical way to construct an isomorphism.

It would be interesting to have an example where two Haar systems give non-isomorphic C^*-algebras.

3. Amenable Groupoids

The notion of amenability for groups (see [41] or [30]) takes many forms and a large part of the theory consists in showing their equivalence. Our goal is much more limited here. We shall first consider measure groupoids and choose a definition of amenability best suited to our needs. We seek a condition ensuring that every representation is weakly contained in the regular representation. Then the von Neumann algebra associated to any representation is injective ; here, the proof is essentially the same as in [83], where R.Zimmer studied ergodic actions of countable discrete groups. A notion of amenability is then given for locally compact groupoids with Haar system, whose main advantage is that it is easily checked.Some examples are studied. Throughout this section, G designates a locally compact groupoid with a fixed Haar system $\{\lambda^u\}$.

3.1. Definition : A quasi-invariant probability measure μ on G^0 will be called amenable (we also say that (G,μ) is amenable) if there exists a net (f_i) in $C_0(G)$ such that

(i) the functions $u \mapsto \int |f_i|^2 d\lambda^u$ converge to 1 in the weak $*$-topology of $L^\infty(G^0,\mu)$ and

(ii) the functions $x \mapsto \int f_i(xy)\ \overline{f_i}(y) d\lambda^{d(x)}(y)$ converge to 1 in the weak $*$-topology of $L^\infty(G,\nu)$ where ν is the induced measure of μ.

This definition reduces to one of the equivalent definitions of amenability in the case of a group ; namely, that the function 1 is the limit, uniformly on compact sets, of functions of the form $f * f^*$, where $f \in C_c(G)$(one has to use theorem 13.5.2 of [19]).

The transitive measures on a principal groupoid provide examples of amenable measures. Let us indicate briefly how the net (f_i) can be constructed. Fix an orbit $[u]$ and let μ be the transitive measure $d_* \lambda^u$. One can choose an increasing net (K_i) of compact sets in $[u]$ (with the topology given by the bijection $d : G^u \to [u]$) such that $\cup K_i = [u]$. Define f_i by

$$f_i(x) = \mu(K_i)^{-1/2} \text{ if } (r,d)(x) \in K_i \times K_i,$$
$$= 0 \text{ otherwise. Then}$$
$$f_i * f_i^*(x) = 1 \text{ if } (r,d)(x) \in K_i \times K_i,$$
$$= 0 \text{ otherwise.}$$

The function f_i is not in $C_c(G)$ but it is in $L^2(G,\nu)$ (where μ has been normalized) and it can be approximated in $L^2(G,\nu)$ by elements of $C_c(G)$.

<u>3.2. Proposition :</u> Let μ be a quasi-invariant amenable probalility measure on G^0 and σ a 2-cocycle in $Z^2(G,\mathbf{T})$. Then the integrated form of any σ-representation of G living on μ is weakly contained in the regular representation on μ of $C^*(G,\sigma)$.

<u>Proof :</u> We follow Takai ([70], page 29). Let $(\mu,\mathcal{K}_u,L)$ be a σ-representation of G. A vector state of the integrated representation is of the form

$$\phi(f) = \int f(y)\ (L(y)\xi\circ d(y),\xi\circ r(y))\ d\lambda_0(y) \text{ for } f \in C_c(G)$$

where ξ is a unit vector in $\Gamma(\mathcal{K})$. Let (f_i) be as in 3.1 and define $\phi_i(f)$ by

$$\phi_i(f) = \int(\int f_i(x)\ \overline{f_i(y^{-1}x)}d\lambda^{r(y)}(x))\quad f(y)\ (L(y)\xi\circ d(y),\xi\circ r(y))d\nu_0(y).$$

By 3.1 (ii), $\phi_i(f)$ tends to $\phi(f)$. Moreover, a routine computation allows us to write the equation

$$\phi_i(f) = \int f(xy)\ (\xi_i(y^{-1}),\ \xi_i(x))\ \sigma(xy,y^{-1})\ d\lambda^u(y)\ d\lambda_u(x)\ d\mu(u)\ ,$$

where $\xi_i(x)$ is defined by

$$\xi_i(x) = D^{1/2}(x)\ \bar{\sigma}(x,x^{-1})\overline{f}_i(x)\ L(x^{-1})\xi\circ r(x).$$

We recognize the expression for $\phi_i(f)$ as $(\mathrm{IndM}(f)\xi_i,\xi_i)$, where IndM is the representation induced by the restriction M of L to $C^*(G^0)$ (see end of 2.7). It acts on the space $\Gamma(\mathcal{K})$ of square-integrable sections of the Hilbert bundle $\mathcal{H}_u = L^2(G,\lambda_u)\otimes\mathcal{K}_u$ on (G^0,μ). Let us compute the norm of ξ_i.

$$\|\xi_i\|^2 = \int \|\xi_i(x)\|^2\ d\nu^{-1}(x)$$
$$= \int |f_i(x)|^2 \|\xi\circ r(x)\|^2\ d\nu(x)\ \text{(because } D = \frac{d\nu}{d\nu^{-1}}\text{)}$$

$$= \int \|\xi(u)\|^2 \left(|f_i(x)|^2 \, d\lambda^u(x)\right) d\mu(u) \,.$$

By 3.1 (i), $\|\xi_i\|$ tends to 1. From the inequality $|\phi_i(f)| \leq \|\mathrm{IndM}(f)\| \; \|\xi_i\| \; \|\xi_i\|$, we obtain $|\phi(f)| \leq \|\mathrm{IndM}(f)\|$. Since IndM is a direct integral of representations equivalent to the regular representation on μ of $C^*(G,\sigma)$, it is weakly contained in it and so is L.

Q.E.D.

3.3. Remark : One expects a converse ; namely, if the integrated form of the trivial one-dimensional representation of G living on μ is weakly contained in the regular representation on μ of $C^*(G)$, then μ is amenable. Let us say that a continuous function ϕ on G is of positive type (with respect to μ) if $\omega_\phi(f) = \int f(x)\, \phi(x)\, d\nu_0(x)$, $f \in C_c(G)$, defines a positive linear functional ω_ϕ on $C^*(G)$. For example, the function 1, which is associated with the vector state $\omega_1(f) = \int f(x) d\nu_0(x)$ of the one-dimensional trivial representation $(\mu, \mathcal{K}_u = \mathbb{C}, L = 1)$ is of positive type. Let us determine the positive type functions associated to the vector states of the regular representation $(\mu, \mathcal{K}_u = L^2(G,\lambda^u), L(x))$ where $L(x)\xi(y) = \xi(x^{-1}y)$. The vector $\xi \in C_c(G) \subset L^2(G,\nu)$ gives the positive type function

$$\begin{aligned}(L(x)\; \xi\circ d(x), \xi\circ r(x)) &= \int \xi(x^{-1}y)\; \overline{\xi}(y) d\lambda^{r(x)}(y) \\ &= \xi * \xi^*(x^{-1}).\end{aligned}$$

Hence, if our hypothesis holds, the state ω_1 is a weak limit of states associated with positive type functions which are finite sums of functions of the form $\xi * \xi^*(x^{-1})$, with $\xi \in C_c(G)$. It is not hard to show that these positive type functions can in fact be chosen to be of the form $\xi * \xi^*(x^{-1})$. Indeed one observes that for ϕ, $f, g \in C_c(G)$,

$$\begin{aligned}\omega_\phi(f * g^*) &= \int \phi(x)\; f * g^*\,(x)\; d\nu_0(x) \\ &= \int \tilde{\phi}(x)\; f * g^*\,(x^{-1})\; d\nu_0(x) \quad (\text{where } \tilde{\phi}(x) = \phi(x^{-1})) \\ &= (L(\tilde{\phi})f, g)\end{aligned}$$

Hence, if ϕ is of positive type, $L(\tilde{\phi})$ is a positive operator. Then, using Kaplansky's density theorem to approximate its square root, one obtains a net (f_i) in $C_c(G)$ such that $L(f_i * f_i^*) \to L(\tilde{\phi})$ in the weak operator topology and $L(f_i * f_i^*) \leq L(\tilde{\phi})$. This implies that the positive linear functionals associated to $f_i * f_i^*\;(x^{-1})$ converge weakly to

ω_ϕ. To conclude, one would need to exhibit ω_1 as a weak limit of states associated with positive type functions of the form $f_i * f_i^* (x^{-1})$ with $f_i \in C_c(G)$ which are uniformly bounded in $L^\infty(G,\nu)$. So far I have been unable to do this.

3.4. Lemma : Let μ be a quasi-invariant probability measure on G^0. Then μ is amenable iff there exists an approximate invariant mean on $L^\infty(G,\nu)$, that is, a net (g_i) of non-negative functions in $C_c(G)$ such that

(i) the functions $u \mapsto \int g_i d\lambda^u$ converge to 1 in the weak $*$-topology of $L^\infty(G^0,\mu)$; and

(ii) the function $x \mapsto \int |g_i(xy) - g_i(y)| d\lambda^{d(x)}(y)$ converge to 0 in the weak $*$-topology of $L^\infty(G,\nu)$.

Proof : The proof is essentially the same as in the case of a group (e.g. [41], page 61). Let us start with (f_i) as in 3.1 and define $g_i = |f_i|^2$. The first property is immediate. Using the inequality $\| a|^2 - |b|^2 | \le (|a| + |b|)(|a - b|)$ and Cauchy-Schwarz, one obtains

$$\int |g_i(xy) - g_i(y)| d\lambda^{d(x)}(y) \le \left[\int (|f_i(xy)| + |f_i(y)|)^2 d\lambda^{d(x)}(y)\right]^{1/2} \left[\int |f_i(xy) - f_i(y)|^2 d\lambda^{d(x)}(y)\right]^{1/2} .$$

Let us set $h_i(u) = \int |f_i(y)|^2 d\lambda^u(y)$. The first member of the product is majorized by

$$2^{1/2} \left[h_i \circ r(x) + h_i \circ d(x)\right]^{1/2}$$

while the second is majorized by

$$\left[|1 - h_i \circ r(x)| + |1 - h_i \circ d(x)| + |1 - f_i * f_i^*(x)| + |1 - f_i * f_i^*(x^{-1})|\right]^{1/2}.$$

The s-topology on $L^\infty(G,\nu)$ is defined by the semi-norms $\alpha_\phi(f) = (\int \phi |f|^2 d\nu)^{1/2}$ where ϕ is a non-negative element of $L^1(G,\nu)$. The first term goes to 2 in the s-topology and is bounded in the $L^\infty(G,\nu)$ norm and the second goes to 0. Thus their product goes to 0 in the s-topology and a fortiori in the weak $*$-topology. Conversely, starting with (g_i), we define $f_i = g_i^{1/2}$. Again, the first property of 3.1 is immediately satisfied. Using the inequality $|a - b|^2 \le |a^2 - b^2|$, one obtains without much trouble the estimate

$$|1 - \int f_i(xy)\overline{f}(y) d\lambda^{d(x)}(y)| \le$$
$$1/2 \left[\int |g_i(xy) - g_i(y)| d\lambda^{d(x)}(y) + |1 - \int g_i(y) d\lambda^{r(x)}(y)| + |1 - \int g_i(y) d\lambda^{d(x)}(y)| \right].$$

Q.E.D.

3.5. Proposition : Let μ be a quasi-invariant amenable probability measure on G^0 and σ a 2-cocycle in $Z^2(G,\mathbb{T})$. Any σ-representation of G living on μ generates an injective von Neumann algebra.

Proof : As mentioned earlier, the idea of the proof is in Zimmer [83]. The notion of amenability we use - it is more stringent than Zimmer's - makes the proof easier. Let $(\mu,\mathcal{K}_u,L)$ be a σ-representation of G ; L also denotes the integrated representation on $\Gamma(\mathcal{K})$ given by $(L(f)\xi,\eta) = \int f(x)(L(x)\xi\circ d(x),\eta\circ r(x))d\nu_0(x)$ $\quad \xi,\eta \in \Gamma(\mathcal{K})$ $\quad f \in C_c(G)$; $\mathcal{M}$ denotes the von Neumann generated by $\{L(f) : f \in C_c(G)\}$; $\mathcal{M}'$ is its commutant and $\mathcal{D}$ is the algebra of decomposable operators on $\Gamma(\mathcal{K})$. An operator $A \in \mathcal{D}$ acts on $\Gamma(\mathcal{K})$ by $A\xi(u) = A(u)\xi(u)$ where $A(u)$ is an operator on $\mathcal{K}_u$. We note that $\mathcal{M}' = \{A \in \mathcal{D} : A\circ r(x)\, L(x) = L(x)\, A\circ d(x)$ for ν a.e. $x\}$. Tomiyama has shown that a von Neumann algebra is injective iff its commutant is injective ; in particular $\mathcal{D}$, which is the commutant of a commutative von Neumann algebra, is injective. We will construct a conditional expectation of $\mathcal{D}$ onto $\mathcal{M}'$; this will show that $\mathcal{M}'$, hence $\mathcal{M}$, is injective. Let (g_i) be a net as in 3.4 and let M be a bound for $\sup_u \int g_i d\lambda^u$. We define a linear map $P_i : \mathcal{D}\to\mathcal{D}$ by

$$P_iB(u) = \int g_i(x)\; L(x)\; B\circ d(x)\; L(x)^{-1} d\lambda^u(x)$$

i.e. $\quad (P_iB\xi,\eta) = \int g_i(x)\; (L(x)B\circ d(x)L(x)^{-1}\xi\circ r(x),\eta\circ r(x))d\nu(x)$

for $\xi,\eta \in \Gamma(\mathcal{K})$.

There is no problem checking that P_i is well defined. Moreover since

$\|P_iB(u)\| \le \|B\| \int g_i(x)d\lambda^u(x)$, we see that

$\|P_iB\| \le M\|B\|$.

We also note that P_i is positive . The P_i's are uniformly bounded in norm. Hence there is a subset converging to a bounded positive linear map P in the following sense. For every pair of vectors (ξ,η) in $\Gamma(\mathcal{K})$ and for every B in $\mathcal{D}$, $(P_iB\xi,\eta)$ tends to $(PB\xi,\eta)$. The restriction of P to $\mathcal{M}'$ is the identity (in particular, P is unital). For if $A \in \mathcal{M}'$, then

$$(P_iA\xi,\eta) = \int g_i(x)\; (A\circ r(x)\; \xi\circ r(x),\eta\circ r(x))d\nu(x)$$
$$= \int (A(u)\xi(u),\eta(u))\; (\int g_i(x)d\lambda^u(x))d\mu(u)$$

By 3.4. (i), we obtain at the limit,

$(PA\xi,\eta) = \int (A(u)\xi(u),\eta(u))d\mu(u) = (A\xi,\eta).$

The proof that P is an expectation will be completed when we show that $P(\mathcal{D}) = \mathcal{M}$.
After routine computations, one obtains, for $B \in \mathcal{D}$, for $f \in C_c(G)$ and $\xi,\eta \in \Gamma(\mathcal{K})$,

$(L(f)P_iB\xi,\eta) =$

$$\int f(x)g_i(y)\sigma(x,y)\bar{\sigma}(y,y^{-1})(L(xy)B\circ d(y)L(y^{-1})\xi\circ d(x),\eta\circ r(x))d\lambda^{d(x)}(y)\, d\nu_0(x), \text{ and}$$

$((P_iB)L(f)\xi,\eta) =$

$$\int f(x)g_i(xy)\sigma(y^{-1}x^{-1},x)\bar{\sigma}(xy,y^{-1}x^{-1})(L(xy)B\circ d(y)L(y^{-1})\xi\circ d(x),\eta\circ r(x))\, d\lambda^{d(x)}(y)\, d\nu_0(x).$$

One notes that

$$\sigma(x,y)\bar{\sigma}(y,y^{-1}) = \sigma(y^{-1}x^{-1},x)\bar{\sigma}(xy,y^{-1}x^{-1}).$$

Hence the following estimate holds:

$$|((L(f)P_iB - P_iBL(f))\xi,\eta)| \le$$

$$\|B\| \int |f(x)|\ \|\xi\circ d(x)\|\ \|\eta\circ r(x)\| \int |g_i(xy) - g_i(y)|\ d\lambda^{d(x)}(y)\, d\nu_0(x).$$

Since

$$\int |f(x)|\ \|\xi\circ d(x)\|\ \|\eta\circ r(x)\|\ d\nu_0(x) \le \|f\|_I\ \|\xi\|\ \|\eta\|,$$

we may use 3.4 (ii) to conclude that the right hand side goes to zero.
Hence $L(f)\,PB = (PB)L(f)$ and $PB \in \mathcal{M}'$.

Q.E.D.

3.5. Remarks :

(a) R.Zimmer has introduced in [82], definition 4.1, the following notion of invariant mean for (G,μ). It is a positive unital linear map m from $L^\infty(G,\nu)$ onto $L^\infty(G^0,\mu)$ satisfying

(i) $m(h\phi) = hm(\phi)$ for $\phi \in L^\infty(G,\nu)$ and $h \in C_c(G^0)$, where $h\phi(x) = h\circ r(x)\phi(x)$, and

(ii) $m(f\phi) = fm(\phi)$ for $\phi \in L^\infty(G,\nu)$, and $f \in C_c(G)$, where $f\phi(x) = \int f(y)\phi(y^{-1}x)d\lambda^{r(x)}(y)$ and where $f\psi(u) = f(y)\psi\circ d(y)d\lambda^u(y)$ for $\psi \in L(G^0,\mu)$. By a compactness argument, the existence of an approximate invariant mean as in 3.4 gives the existence of an invariant mean. The converse is probably true, but I don't have a correct proof. It can be shown as in [82] and [83], where the case of an ergodic action of a countable discrete group is considered, that for a discrete groupoid G, the regular representation on μ generates an injective von Neumann algebra iff there is an invariant mean for (G,μ).

(b) R. Zimmer has also defined in [81], definition 1.4, an amenable ergodic group action by a fixed point property. This property is equivalent to the existence of an invariant mean in the discrete case ([82], 4.1) but the general case is unknown. The definition of amenability given in 3.4 implies the fixed point property, as a standard averaging process shows.

(c) One can also use the approximate invariant mean of 3.4 to average cocycles and get a vanishing theorem (cf. Johnson, [48], 2.5, page 32).

3.6. Definition : Let us say that G is measurewise amenable if every quasi-invariant measure on G^0 is amenable.

If all the representations of $C^*(G,\sigma)$ are obtained by integration and if G is measurewise amenable, it results from 3.2 that $C^*(G,\sigma)$ coincides with the reduced C^*-algebra $C^*_{red}(G,\sigma)$ and from 3.5 that it is nuclear.

A sufficient condition for G to be measurewise amenable is the existence of a net (f_i) in $C_c(G)$ such that

(i) the functions $u \mapsto \int |f_i(x)|^2 d\lambda^u(x)$ are uniformly bounded in the sup-norm ; and

(ii) the functions $x \mapsto \int f_i(xy)\overline{f}_i(y)d\lambda^{d(x)}(y)$ converge to 1 uniformly on any compact subset of G. This condition is also necessary in the case of a group (cf. Dixmier [19], 13.5.2, page 260) ; but I do not know if it is true in general. Since this condition is handy, I call it amenability, although I don't have any real justification for it.

A question which arises is the amenability of $C^*(G,\sigma)$ in the sense of Johnson ([48], 5, page 60) ; in particular, does the above condition imply amenability ?

Let us now look at how amenability is preserved under some operations.

3.7. Proposition : Let U be a locally closed subset of the unit space of G. If G is [measure wise] amenable, the reduction G_U is [measure wise] amenable.

Proof : Suppose G amenable. Then there exists a net (f_i) in $C_c(G)$ such that $f_i * f_i^*$ converges to 1 uniformly on the compact sets of G and $|f_i * f_i^*(u)| \leq M$ for suitable M and any u. Let (h_i) be an approximate identity on $C_c(U)$, bounded in sup-norm. Then

$g_i(x) = h_i \circ r(x)\ f_i(x)\ h_i \circ d(x),\ x \in G_U$

defines a net in $C_c(G_U)$ satisfying the required properties. The proof of the other statement is similar. We note that any quasi-invariant on U is equivalent to the restriction to U of a quasi-invariant measure on G^0, namely, a quasi-invariant measure μ on U is equivalent to the restriction of the saturation $[\mu]$ of μ with respect to G (1.3.7). We denote the restriction of λ^u to G_U by λ^u_U and the induced measure with respect to $\{\lambda^u_U\}$ by $\nu_U = \int \lambda^u_U\, d\mu(u)$. Then for $E \subset U$,

$[\mu](E) = 0$ iff $\nu(d^{-1}(E)) = 0$;

iff for μ a.e. u, $\lambda^u(d^{-1}(E)) = 0$;

iff for μ a.e. u, $\lambda^u_U(d^{-1}(E)) = 0$ (because E U and μ lives on U) ;

iff $\nu_U(d^{-1}(E)) = 0$;

iff $\nu_U^{-1}(d^{-1}(E)) = 0$ (because μ is quasi-invariant) ;

iff $\mu(E) = 0$.

Q.E.D.

<u>3.8. Proposition :</u> Let G be a locally compact groupoid with Haar system, let A be a locally compact group and c a continuous 1-cocycle in $Z^1(G,A)$. Let G(c) be their skew product (1.1.6).

(i) If G is [measurewise] amenable, then G(c) is [measurewise] amenable.

(ii) If A is amenable and G(c) is [measurewise] amenable, then G is [measurewise] amenable.

<u>Proof :</u> Let us recall the definition 1.1.6 of G(c) : $G(c) = G \times A$ with $(x,a)(y,ac(x)) = (xy,a)$ and $(x,a)^{-1} = (x^{-1},ac(x))$. Its unit space is $G^0 \times A$. The locally compact groupoid G(c) has been defined before 1.4.10. If $\{\lambda^u\}$ is a Haar system for G, a Haar system $\{\lambda^{u,a}\}$ for G(c) is given by

$$\int f(x,b)d\lambda^{u,a}(x,b) = \int f(x,a)d\lambda^u(x) .$$

Let us describe quasi-invariant measures for G(c). Suppose that $\underline{\mu}$ is a quasi-invariant measure on G^0 for G and $\{\alpha_u\}$ a system of measures on A which is $\underline{\mu}$-adequate (Bourbaki [6] 3.1) (this means that $\mu = \int \alpha_u d\underline{\mu}(u)$ is well defined) and which satisfies $\alpha_{d(x)} \sim \alpha_{r(x)}c(x)$ for $\underline{\nu}$ a.e. x, where $\underline{\nu}$ is the induced measure on $\underline{\mu}$. Then $\mu = \int \alpha_u d\mu(u)$

is a quasi-invariant measure for $G(c)$. Conversely, if the quasi-invariant measure μ can be disintegrated along the first projection of $G^0 \times A$, it is of that form. For the proof of this fact, we may assume that $\mu, \underline{\mu}$ and $\underline{\mu}$ a.e. α_u are probability measures and we may replace $\{\lambda^u\}$ by equivalent probability measures. The measure ν induced by μ is of the form

$$\int f d\nu = \int f(x,a) d\lambda^u(x) d\alpha_u(a) d\underline{\mu}(u),$$

while ν^{-1} is of the form

$$\int f d\nu^{-1} = \int f(x^{-1}, ac(x)) d\lambda^u(x) d\alpha_u(a) d\underline{\mu}(u) .$$

In particular, for any measurable set E in C, $\nu(E \times A) = \underline{\nu}(E)$ and $\nu^{-1}(E \times A) = \underline{\nu}^{-1}(E)$. This shows that $\underline{\mu}$ is quasi-invariant. Then, using the uniqueness of the disintegration of ν along the first projection of $G \times A$, one gets $\alpha_{r(x)} \sim \alpha_{d(x)}\, c(x^{-1})$ for ν a.e. x. Conversely, there is no problem checking that a measure μ of the above form is quasi-invariant.

(i) Let $\mu = \int \alpha_u d\underline{\mu}(u)$ be a quasi-invariant measure for $G(c)$ as above. If $\underline{\mu}$ is amenable, there exists a net (f_i) in $C_c(G)$ such that $u \to f_i * f_i^*(u)$ converges to 1 weakly $*$ in $L^\infty(G^0, \underline{\mu})$, and $x \to f_i * f_i^*(x)$ converges to 1 weakly $*$ in $L^\infty(G, \underline{\nu})$. Let (h_i) be an approximate identity for $C_c(A)$ with pointwise multiplication and bounded in sup-norm and define $g_i \in C_c(G \times A)$ by

$$g_i(x,a) = f_i(x) h_i(a).$$

The net (g_i) has the required properties :

$$g_i * g_i^*(x,a) = \int g_i(xy,a)\, \bar{g}_i(y, ac(x))\, d\lambda^{d(x)}(y)$$
$$= h_i(a)\, \bar{h}_i(ac(x))\, f_i * f_i^*(x).$$

Let us check the convergence of $(u,a) \to g_i * g_i^*(u,a)$. The net is bounded in $L^\infty(G^0 \times A, \mu)$, hence it is enough to check the convergence against functions of the form $f(u)g(a)$ where $f \in C_c(G^0)$ and $g \in C_c(A)$.

We see that

$$\int f(u) g(a)\, g_i * g_i^*(u,a)\, d\mu(u,a)$$
$$= \int f(u) \left(\int g(a) |h_i(a)|^2 d\alpha_u(a)\right) f_i * f_i^*(u)\, d\mu(u) \text{ goes to}$$
$$\int f(u)\, g(a)\, d\mu(u,a),$$

Since $\int g(a) |h_i(a)|^2 d\alpha_u(a)$ goes to $\int g(a) d\alpha_u(a)$ in $L^1(G^0, \mu)$ and $f_i * f_i^*(u)$ goes to 1 in $(L^\infty(G^0, \mu)$, weak$*)$. The convergence of $g_i * g_i^*(x,a)$ is proved in the same fashion.

This shows that μ is amenable. One proves in the same way that if G is amenable, then G(c) is amenable.

(ii) We assume that A is amenable and G(c) is measurewise amenable. Let $\underline{\mu}$ be a quasi-invariant measure on G^0. Then $\mu = \underline{\mu} \times \alpha$, where α is a right Haar measure for A, is quasi-invariant for G(c). Since G(c) is amenable, there exists an approximate invariant mean (g_i), $g_i \geq 0$, $g_i \in C_c(G \times A)$ such that $(u,a) \to \int g_i(x,a)d\lambda^u(x)$ converges to 1 in $(L^\infty(G^0 \times A,\mu)$, weak*), and $(x,a) \to \int |g_i(xy,a) - g_i(y,ac(x)| \;\; d\lambda^{d(x)}(y)$ converges to 0 in $(L^\infty(G \times A,\nu)$, weak *). The group A also has an approximate invariant mean (k_i) : $k_i \geq 0$ $k_i \in C_c(A)$ such that $\int k_j(a)d\alpha(a) = 1$, and $b \to \int |k_j(ab) - k_u(a)| \; d\alpha(a)$ converges to 0 uniformly on the compact subsets of A. Let us define $f_{ij} \in C_c(C)$ by $f_{ij}(x) = \int g_i(x,a)k_j(a)d\alpha(a)$. It is not hard to check that the family of functions $u \mapsto \int f_{ij}(x)d\lambda^u(x)$ is bounded in $L^\infty(G^0,\underline{\mu})$ and the family of functions $x \mapsto \int |f_{ij}(xy) - f_{ij}(y)|d\lambda^{d(x)}(y)$ is bounded in $L^\infty(G,\underline{\nu})$. We will show that, given a neighborhood of 1 in $(L^\infty(G^0,\underline{\mu})$, weak*),

$$V = \{h \in L^\infty(G^0,\underline{\mu}) : \;\; |\int(h(u) - 1) \; \phi_k(u) \; d\underline{\mu}(u)| \leq \varepsilon_k, \; k=1,\dots,m\},$$

where $\phi_k \in C_c(G^0)$, $\varepsilon_k > 0$, $k=1,\dots,m$, and a neighborhood of 0 in $(L^\infty(G,\underline{\nu})$, weak*),

$$W = \{f \in L^\infty(G,\underline{\nu}) : | \int f(x)\psi_\ell(x) \;\; d\underline{\nu}(x)| \leq \eta_\ell \;\; \ell=1,\dots,n\} ,$$

there exists f_{ij} such that $u \to \int f_{ij} \; d\lambda^u$ is in V and $x \to \int |f_{ij}(xy) - f_{ij}(y)|d\lambda^{d(x)}(y)$ is in W. Let M be a bound for the norm of the functions $(u,a) \to \int g_i(x,a) \; d\lambda^u(x)$ in $L^\infty(G^0 \times A)$. We can choose j such that, for every $\ell = 1,\dots,n$,

$$\int |\psi_\ell(x)| \; |k_j(ac(x)^{-1}) - k_j(a)| \;\; d\alpha(a) \; d\underline{\nu}(x) \;\; \leq \eta_\ell/2M$$

from now on, j is kept fixed. We observe that

$$\int \phi_k(u) \; (\int f_{ij}(x) \; d\lambda^u(x)) \; d\underline{\mu}(u)$$
$$= \int (g_i(x,a) \; d\lambda^u(x)) \; \phi_k(u) \; k_j(a) \; d\mu(u,a)$$

goes to $\int \phi_k(u) \; k_u(a) \; d\mu(u,a) = \int \phi_k(u) \; d\mu(u)$ as i goes to ∞. Hence for i sufficiently large, $u \to \int f_{ij}d\lambda^u$ is in V. Similarly, for i sufficiently large, we will have

$$\int (\int |g_i(xy,a) - g_i(y,ac(x)) | \; d\lambda^{d(x)}(y)) \; |\psi_\ell(x)|k_j(a)d\nu(x,a) \; \leq \eta_\ell \; /2$$
$$\ell = 1,\dots,n \; .$$

Writing

$$f_{ij}(xy) - f_{ij}(y)$$
$$= \int(g_i(xy,a) - g_i(y,ac(x)))k_j(a)d\alpha(a) + \int g_i(y,a)(k_j(ac(x)^{-1}) - k_j(a)) \; d\alpha(a),$$

we obtain the estimate

$$|\int(\int|f_{ij}(xy) - f_{ij}(y)|d\lambda^{d(x)}(y))\ \psi_\ell(x)\ d\underline{\nu}(x)|$$
$$\leq \int(\int|g_i(xy,a) - g_i(y,ac(x))|d\lambda^{d(x)}(y)\ |\psi_\ell(x)|k_j(a)\ d\nu(x,a)$$
$$+\int|\psi_\ell(x)|\ \ (\int g_i(y,a)d\lambda^{d(x)}(y))\ |k_j(ac(x)^{-1})-k_j(a)|\ \ d\nu(x,a)$$
$$\leq \eta_\ell$$

This shows that $\underline{\mu}$ is amenable. One proves in the same way that the amenability of G(c) and A implies the amenability of G.

Q.E.D.

Dual statements hold for the semi-direct product.

<u>3.9: Proposition :</u> Let G be a locally compact groupoid with Haar system, let A be a locally compact group acting continuously on G by automorphisms leaving the Haar system invariant and let $G \times_\alpha A$ be their semi-direct product (1,1.7).

(i) If A is amenable and G is [measurewise] amenable, then $G \times_\alpha A$ is [measurewise] amenable.

(ii) If the semi-direct product $G \times_\alpha A$ is [measurewise] amenable, then G is [measurewise] amenable.

<u>Proof :</u> Let us first define the semi-direct product as a locally compact groupoid with Haar system. We require the map from A x G into G sending (a,x) into s(a) x to be continuous. Recall (1.1.7) that $G \times_\alpha A$ is the groupoid G x A with $(x,a)(y,b) = (x(s(a)y),\ ab)$ and $(x,a)^{-1} = (s(a^{-1})x^{-1},a^{-1})$. Its unit space may be identified with G^0. The product topology makes it into a locally compact groupoid. We say that the automorphism s of G leaves the Haar system invariant if $s\cdot\lambda^u = \lambda^{s(u)}$; in other words $\int f(s(x))d\lambda^{s^{-1}(u)}(x) = \int f(x)d\lambda^u(x)$ for $f \in C_c(G^u)$. If $\{\lambda^u\}$ is a Haar system for G and α a left Haar measure for A, then $\{\lambda^u \times \alpha\}$ is a Haar system for $G \times_\alpha A$. Let us check left invariance :

$$\int f(x,a)(y,b))d\lambda^{s(a^{-1})d(x)}(y)d\alpha(b)$$
$$= \int f(x(s(a)y),ab)d\lambda^{s(a^{-1})d(x)}(y)d\alpha(b)$$
$$= \int f(xy,ab)d\lambda^{d(x)}(y)d\alpha(b) = \int f(y,b)d\lambda^{r(x)}(y)d\alpha(b).$$

Since the proof of this proposition is not much different from the previous one and does not involve any difficulty, we will just indicate how the various approximate means may be constructed.

(i) Given (f_i) such that $f_i * f_i^* \to 1$ in G and (h_j) such that $h_j * h_j^* \to 1$ on A, set

$$g_{ij}(x,a) = f_i(s(a^{-1})x)h_j(a).$$

Then, there exists a subnet such that $g_{ij} * g_{ij}^* \to 1$ in $G \times_\alpha A$.

(ii) Given an approximate invariant mean (g_i) for $G \times_\alpha A$, we can define an approximate invariant mean for G by setting

$$f_i(x) = \int g_i(x,a)\, d\alpha(a).$$

Q.E.D.

3.10. Example : A transformation group arising from the action of an amenable group is always amenable but the converse is not true. Let G be a second countable locally compact group and H a closed subgroup ; it can be shown that the transformation group $H\backslash G \times G$ is amenable iff H is amenable. Hence a homomorphic image of an amenable groupoid is not necessarily amenable ; however, it is probably true that the asymptotic range (1.4.3) of such a homomorphism is amenable (cf. Zimmer [81], 3.3).

In conclusion, let us ask some very basic questions.

(i) Is a closed subgroupoid of a measurewise amenable groupoid also measurewise amenable ? This is probably true but I can prove it only in the case of an r-discrete groupoid. The proof uses 3.3 and 4.1.

(ii) Does 3.9 hold for more general extensions ?

(iii) Is amenability preserved under (the appropriate notion of) similarity ?

4. The C^*-Algebra of an r-Discrete Principal Groupoid

Reduced C^*-algebras of r-discrete principal groupoids are generalizations in all essential respects of the usual $*$-algebras of matrices. They appear in a diagonalized form. That is, $C^*(G^0)$ is a maximal abelian subalgebra, the image of a unique conditional expectation. The elements of $C^*_{red}(G,\sigma)$ are matrices over G, the diagonal matrices are the elements of $C^*(G^0)$ and the expectation map is evaluation on the diagonal. The ideal structure of $C^*_{red}(G,\sigma)$ is easily described. Ideals correspond to open invariant subsets of the unit space. Part of the representation theory may be conveniently expressed in terms of the groupoid. For example, the regular representation on μ is primary [resp. type I, II or III] iff the measure μ is ergodic [resp. type I, II or III]. Such C^*-algebras are characterized by the existence of a special kind of maximal abelian subalgebras, which, in accordance with [31], where a similar notion is introduced in the context of von Neumann algebras, we call Cartan subalgebras.

In the following proposition, we use the reduced norm $\| \ \|_{red}$, which has been defined in 2.8 and the sup-norm $\| \ \|_\infty$.

4.1. Proposition : Let G be an r-discrete groupoid with Haar system and let σ be a continuous 2-cocycle. Then, the following inequalities hold for any $f \in C_c(G,\sigma)$:

(i) $\|f\|_\infty \leq \|f\|_{red}$; and

(ii) for any $u \in G^0$, $\int |f|^2 d\lambda_u \leq \|f\|^2_{red}$.

The proof results directly from the following lemma.

Lemma : Let G and σ be as above and let x be a point in G with $d(x) = u$. Consider the representation L of $C_c(G,\sigma)$ induced by the point mass at u (see 2.7). Let ξ and η be the unit vectors δ_u and δ_x respectively in the space $L^2(G,\lambda_u)$ of the representation L. Then for any $f \in C_c(G,\sigma)$, $f(x) = (L(f)\xi,\eta)$ and $f(y) = L(f)\xi(y)$ for any $y \in G_u$.

Proof : This is immediate since L is given by $(L(f)\xi,\eta) =$ $\int f(yz)\xi(z^{-1})\,\bar{\eta}(y)\,\sigma(yz,z^{-1})d\lambda^u(z)d\lambda_u(y)$ (see 2.7). Note also that, since G is r-discrete, λ_u is the counting measure on G_u(see 1.2.7). For the proof of the proposition, note that $|f(x)| \leq \|L(f)\| \ \|\xi\| \ \|\eta\| < \|f\|_{red}$ by definition 2.8, and so $\int |f(y)|^2 d\lambda_u(y) = \|L(f)\xi\|^2 \leq \|L(f)\|^2 \ \|\xi\|^2 \ \leq \|f\|_{red}$.

Q.E.D.

The injection j of $C_c(G)$ into $C_0(G)$, the Banach space of continuous functions on G which vanish at infinity, extends to a norm decreasing linear map j of $C^*_{red}(G,\sigma)$ into $C_0(G)$.

<u>4.2. Proposition :</u> Let G be an r-discrete groupoid with Haar system and σ a continuous 2-cocycle. Then

(i) the map j from $C^*_{red}(G,\sigma)$ to $C_0(G)$ is one-to-one (therefore, the elements of $C^*_{red}(G,\sigma)$ will be viewed as functions on G) ;

(ii) any $a \in C^*_{red}(G,\sigma)$ satisfies $\|a\|_\infty \leq \|a\|_{red}$ and $\| \, |a|^2 \, \|_I \leq \|a\|^2_{red} \leq \|a\|^2_I$, where the norm $\| \ \|_I$ has been defined in 1.4 ($\|a\|_I$ may be infinite) ; and

(iii) the operations in the $*$-algebra $C^*_{red}(G,\sigma)$ may be expressed in the same way as in the same way as in the $*$-algebra $C_c(G,\sigma)$, explicitly

$a^*(x) = \bar{a}(x^{-1})\bar{\sigma}(x,x^{-1})$, for $a \in C^*_{red}(G,\sigma)$,

$a * b(x) = \int a(xy)b(y^{-1})\,\sigma(xy,y^{-1})d\lambda^{d(x)}(y)$, for $a,b \in C^*_{red}(G,\sigma)$, and

$ha(x) = h\circ r(x)\, a(x)$, for $h \in C^*(G^0)$ and $a \in C^*_{red}(G,\sigma)$.

<u>Proof :</u>

(i) Let μ be a quasi-invariant probability measure on G^0. The regular representation on μ is realized in standard form on $L^2(G,\nu^{-1})$ (see 1.10). This representation is the GNS representation associated with the state $\mu\circ P(f) = \int P(f)d\mu = (L(f)\phi_0,\phi_0)$ where P is the restriction map $C_c(G,\sigma) \to C_c(G^0)$ and ϕ_0 is the characteristic function of G^0, considered as a unit vector in $L^2(G,\nu^{-1})$. In particular, ϕ_0 is cyclic and separating for the left representation. We may write $L(f)\phi_0 = f * \phi_0 = \underline{j}(f)$ for $f \in C_c(G,\sigma)$ where $\underline{j}$ is the natural map from $C_c(G)$ into $L^2(G\,\nu^{-1})$. We note that by 4.1 $\|j(f)\| \leq \|f\|_{red}$. Hence the equality remains true for $a \in C^*_{red}(G,\sigma)$, $L(a)\phi_0 = \underline{j}(a)$. As $\underline{j}(a) = j(a)\,\nu^{-1}$ a.e., $j(a) = 0 \Longrightarrow \underline{j}(a) = 0 \Longrightarrow L(a) = 0$. Since the regular representations form a faithful family of representations of $C^*_{red}(G,\sigma)$, $a = 0$.

(ii) By continuity, the inequalities of 4.1 still hold for $a \in C^*_{red}(G,\sigma)$. The inequality $\|a\|_{red} \leq \|a\|_I$ has been written here for completeness.

(iii) It suffices to justify the passage to the limit in the expressions which are valid for $f \in C_c(G,\sigma)$. For example, suppose that $f_n \to a$ and $g_n \to b$ in $C^*_{red}(G,\sigma)$, with $f_n, g_n \in C_c(G,\sigma)$. Then $f_n * g_n(x) = \int f_n(xy)g_n(y^{-1})\sigma(xy,y^{-1})d\lambda^{d(x)}(y)$. Because of the estimate $\| \ \|_\infty \leq \| \ \|_{red}$, $f_n * g_n(x) \to a * b(x)$ ($a * b$ denotes the product of a and b). On the other hand, because of the estimate $\| \ \|_2 \leq \| \ \|_{red}$ where $\| \ \|_2$ is the norm of $L^2(G,\lambda^{d(x)})$, $f_n(x\cdot) \to a(x\cdot)$ and $g_n \to b$ in $L^2(G,\lambda^{d(x)})$, hence the right hand side goes to $\int a(xy)b(y^{-1})\sigma(xy,y^{-1})d\lambda^{d(x)}(y)$.

Q.E.D.

Remark : (Cf. [31,II]). It seems hard to characterize the range of the map j. In the case of the principal groupoid I × I, where I is a countable discrete space, this amounts to characterizing the matrices of compact operators. We may note that, in this case, the conditions $\int |a(x)|^2 \, d\lambda^u(x) \leq \|a\|^2$ are satisfied by the matrix of any bounded operator.

Let us study now the ideal structure of the reduced C^*-algebra of an r-discrete principal groupoid. In fact one can do a little better.

4.3. Definition : Let us say that a locally compact groupoid G is essentially principal when for every invariant closed subset F of its unit space, the set of u's in F whose isotropy group G(u) is reduced to {u} is dense in F.

4.4. Proposition : Let G be an r-discrete essentially principal groupoid with Haar system. Then for any quasi-invariant measure μ, any σ-representation L of G on μ and any $f \in C_c(G)$, the following inequality holds :

$$\sup_{u \in F} |f(u)| < \|L(f)\| \quad \text{where F is the support of } \mu.$$

Proof :

It suffices to prove the inequality $|f(\underline{u})| < \|L(f)\|$ for $\underline{u} \in F$ such that $G(\underline{u}) = \{\underline{u}\}$.

Let (V_n) be a fundamental sequence of neighborhoods of $\underline{u}$.

There exists a sequence (ξ_n) of square-integrable sections of the Hilbert bundle of the representation satisfying

$$\operatorname{Supp} \xi_n \subset V_n \quad \text{and} \quad \int |\xi_n(u)|^2 \, d\mu(u) = 1$$

We will show that the sequence $(L(f)\xi_n, \xi_n)$ tends to $f(\underline{u})$.

We first write f as a finite sum of functions supported on compact open G-sets :

$$f = \sum_1^m f_i \quad , \quad f_i = h_i \chi_{S_i} \quad \text{with } h_i \in C_c(G^0) \text{ and } S_i \in \mathbf{G}.$$

We use 1.7. (*) to compute $(L(f_i)\xi_n, \xi_n)$:

$$(L(f_i)\xi_n,\xi_n) = \int_{V_n \cap V_n \cdot S_i^{-1}} h_i(u)(L(u\, S_i)\xi_n(u \cdot S_i),\xi_n(u))\, D^{-1/2}(u\, S_i)\, d\mu(u)$$

If $\underline{u} \neq \underline{u} \cdot S_i$, we have eventually $V_n \cap V_n \cdot S_i^{-1} = \emptyset$ and $(L(f_i)\xi_n, \xi_n) = 0$.

If $\underline{u} = \underline{u} \cdot S_i$, the G-set S_i meets G^0. For n large enough, $V_n S_i$ is contained in G^0 and $(L(f_i)\xi_n, \xi_n) = \int_{V_n} h_i(u) \, \|\xi_n(u)\|^2 \, d\mu(u)$ tends to $h_i(\underline{u})$.

Q.E.D.

Let G be an arbitrary locally compact groupoid with Haar system. Its reduced C^*-algebra has a distinguished family of ideals, defined by invariant open subsets of G^0. This is well known in the cases studied previously (e.g. [86], 2.29). Let us introduce some notation. $\mathcal{J}(A)$ will denote the lattice of ideals of the C^*-algebra A. $\mathcal{O}(G)$ will denote the lattice of invariant open subsets of the unit space of the groupoid G.

For U in $\mathcal{O}(G)$, $I_c(U) = \{f \in C_c(G,\sigma) : f(x) = 0 \text{ if } x \notin G_U\}$ and $I(U)$ is the closure of $I_c(U)$ in $C^*_{red}(G,\sigma)$.

<u>Lemma :</u> Let X be a locally compact space and Y be a normal open subspace. Then, the closure of $\{f \in C_c(X) : \text{supp} f \subset Y\}$ in the inductive limit topology of $C_c(X)$ is $\{f \in C_c(X) : f(x) = 0 \text{ if } x \notin Y\}$.

<u>Proof :</u> One constructs an approximate identity for $C_c(Y)$ as follows. There are increasing nets (V_α) and (V'_α) of relatively compact open subsets of Y such that $\bar{V}_\alpha \subset V'_\alpha$ while $\cup V_\alpha = \cup V'_\alpha = Y$ and there are functions $e_\alpha \in C_c(Y)$ supported on V'_α such that $e_\alpha = 1$ on V_α. If $f \in C_c(X)$ and $f(x) = 0$ if $x \notin Y$, $\text{supp}(fe_\alpha) \subset Y$ and $fe_\alpha \to f$ in the inductive limit topology. Moreover, the set $\{f \in C_c(X) : f(x) = 0 \text{ if } x \notin Y\}$ is clearly closed.

<u>4.5. Proposition :</u> Let G be a locally compact groupoid with Haar system and let σ be a continuous 2 cocycle.

(i) If U is an invariant open subset of G^0 and F is its complement, then $I(U)$ is an ideal of $C^*_{red}(G,\sigma)$ which is isomorphic to $C^*_{red}(G_U,\sigma)$ and such that the quotient is isomorphic to $C^*_{red}(G_F,\sigma)$.

(ii) If μ be a quasi-invariant measure of support F, the ideal $I(U)$, where U is the complement of F, is the kernel of the regular representation on μ.

(iii) The correspondence $U \mapsto I(U)$ is a one-to-one order preserving map from $\mathcal{O}(G)$ into $\mathcal{J}(C^*_{red}(G,\sigma))$.

Proof :

(i) Using the invariance of U, one easily checks that $I_c(U)$ is a self-adjoint two-sided ideal of $C_c(G,\sigma)$. Indeed, suppose $f \in I_c(U)$ and $g \in C_c(G,\sigma)$, then

$$f * g\,(x) = \int f(y)g(y^{-1}x)\sigma(y,y^{-1}x)\, d\lambda^{r(x)}(y).$$

If $x \notin G_U$ and $r(y) = r(x)$, $y \notin G_U$ and $f(y) = 0$, hence $f * g(x) = 0$. Therefore its closure in $C^*_{red}(G,\sigma)$ is a closed ideal of $C^*_{red}(G,\sigma)$. The map j from $C_c(G_U,\sigma)$ to $C_c(G,\sigma)$ which extends a function on G_U by 0 outside of G_U is a $*$-homomorphism and is isometric for the reduced norm. In fact, if we compose it with the regular representation on μ, where μ is a quasi-invariant measure on G, we obtain the regular representation on μ_U, the restriction of μ to U. Conversely, if μ is a quasi-invariant measure on U, it can be viewed as a quasi-invariant measure $\underline{\mu}$ on G^0 and

$\mathrm{Ind}\mu\ (f) = \mathrm{Ind}\underline{\mu}\ (j(f))$ for $f \in C_c(G_U,\sigma)$. Hence we have an isometric $*$-homomorphism from $C^*_{red}(G_U,\sigma)$ to $C^*_{red}(G,\sigma)$. The lemma shows that its image is $I(U)$.

The restriction map p from $C_c(G,\sigma)$ onto $C_c(G_F,\sigma)$ is a $*$-homomorphism. If μ is a quasi-invariant measure on F, we view it as a quasi-invariant measure on G^0, say $\underline{\mu}$. We have $\mathrm{Ind}\mu\ (p(f)) = \mathrm{Ind}\underline{\mu}\ (f)$ for $f \in C_c(G,\sigma)$. Hence p decreases the reduced norm and extends to a $*$-homomorphism from $C^*_{red}(G,\sigma)$ onto $C^*_{red}(G_F,\sigma)$. Its kernel I clearly contains $I(U)$. Let L be a representation of $C^*_{red}(G,\sigma)$ which vanishes on $I(U)$. We define L_F on $C_c(G_F,\sigma)$ by $L_F(f) = L(f')$ where $f' \in C_c(G,\sigma)$ and $f'_{|G_F} = f$. This makes sense because L vanishes on $I_c(U)$. The map L_F is a representation of $C_c(G_F,\sigma)$ and satisfies $L_F \circ p\ (f) = L(f)$. If μ_1 and μ_2 are disjoint quasi-invariant measures on G^0, $\mathrm{Ind}\mu_1$ and $\mathrm{Ind}\mu_2$ are disjoint representations and $\|\mathrm{Ind}_{\mu_1 \vee \mu_2}(f')\| = \mathrm{Max}(\|\mathrm{Ind}\mu_1(f')\|, \|\mathrm{Ind}\mu_2(f')\|)$. This gives the estimate $\|L_F(f)\| < \|L\|\ \|f\|_{red}$. Hence L_F extends to a representation of $C^*_{red}(G_F,\sigma)$ and L factors through p. Therefore $I = I(U)$.

(ii) It suffices to show that the regular representation on a quasi-invariant measure μ of support G^0 is a faithful representation of $C^*_{red}(G,\sigma)$. Let M be a representation of G^0. It is weakly contained in the representation defined by μ. Since the process of induction preserves weak containment, the kernel of $\mathrm{Ind}M$ contains the kernel of $\mathrm{Ind}\mu$.

(iii) This is clear.

Q.E.D.

We are ready to give the announced result on the ideal structure of the reduced C^*-algebra of a principal r-discrete groupoid. It is well known in the case of a transformation group (corollary 5.16 of [24] and theorem 5.15 of [86]). Without the assumption of r-discreteness, the problem has recently been solved, in the case of a transformation group, by E.Gootman and J.Rosenberg [38].

4.6. Proposition : Let G be an r-discrete essentially principal groupoid with Haar system and σ a continous 2-cocycle. Then the correspondence $U \mapsto I(U)$ is an order preserving bijection between the lattice $\mathcal{O}(G)$ of invariant open subsets of G^0 and the lattice $\mathcal{J}(C^*_{red}(G,\sigma))$ of ideals of the reduced C^*-algebra $C^*_{red}(G,\sigma)$.

Proof : Let L be a σ-representation of G living on the quasi-invariant probability measure μ. Let P be the restriction map from $C_c(G,\sigma)$ onto $C_c(G^0)$. We have seen (comments before 2.7) that P is a conditional expectation and that $\mathrm{Ind}\mu$ is the GNS representation associated with the state $\mu \circ P$. It results from 4.4 that $|\mu\circ P(f)| \leqslant \|L(f)\|$, hence $\|\mathrm{Ind}\mu\ (f)\| \leqslant \|L(f)\|$ for any $f \in C_c(G)$.

In particular, if L is a representation of $C^*_{red}(G,\sigma)$, its kernel is contained in $I(U)$ where U is the complement of the support of μ. Since the reverse inclusion is clear, its kernel is precisely $I(U)$. Hence the map $U \mapsto I(U)$ is onto.

Q.E.D.

Our next task is to justify the statement that the reduced C^*-algebra of an r-discrete principal groupoid appears in a diagonalized form.

4.7. Proposition : Let G be an r-discrete groupoid with Haar system and σ a continuous 2-cocycle. Then

(i) an element a of $C^*_{red}(G,\sigma)$ commutes with every element of $C^*(G^0)$ iff it vanishes off the isotropy group bundle $G' = \{x \in G : d(x) = r(x)\}$; and

(ii) $C^*(G^0)$ is a maximal subalgebra of $C^*_{red}(G,\sigma)$ iff G^0 is the interior of G'.

Proof : Since G^0 is open, $C_c(G^0)$ is a subalgebra of $C_c(G,\sigma)$ and $C^*(G^0)$ is a subalgebra of $C^*_{red}(G,\sigma)$. It consists exactly of those elements of $C^*_{red}(G,\sigma)$ which vanish off the unit space G^0. Let $a \in C^*_{red}(G,\sigma)$ and $h \in C^*(G^0)$. Then $ah(x) = a(x)h\circ d(x)$ and $ha(x) = h\circ r(x)\, a(x)$. If $a(x) = 0$ for any x such that $d(x) \neq r(x)$, then $a(x)h\circ d(x) = h\circ r(x)a(x)$ holds for every x in G. If $a(x) \neq 0$ for some x such that $d(x) \neq r(x)$, then there exists $h \in C^*(G^0)$ such that $h\circ d(x) = 1$ and $h\circ r(x) = 0$, consequently, $a(x)h\circ d(x) \neq h\circ r(x)a(x)$, and so $ah \neq ha$. The assertion (ii) is an immediate consequence of (i).

Q.E.D.

<u>4.8. Proposition :</u> Let G be an r-discrete principal groupoid with Haar system and σ a continuous 2-cocycle. Then the restriction map $P : C^*_{red}(G,\sigma) \to C^*(G^0)$ is the unique conditional expectation onto $C^*(G^0)$ and is faithful.

<u>Proof :</u> The proof is identical to a proof one would give in the case of matrix algebras. Note that, by 4.2, P is well defined. There is no difficulty checking P has all the properties of an expectation map. To show uniqueness, we use the same device as in 4.3 or 4.5.d. Let $a \in C_c(G,\sigma)$ and suppose that suppa does not meet the diagonal Δ of $G^0 \times G^0$ - again, we view G as a subset of $G^0 \times G^0$. There exists a finite cover of r(supp a) by open sets U_i i=1,...,n on G^0 such that $U_i \times U_i \cap \text{supp } a = \emptyset$ for i=1,...,n. Let (h_i) be a partition of unity subordinate to this cover, with $\sum h_i(u) = 1$ for $u \in r(\text{supp } a)$. Then $a = (\sum_1^n h_i)\, a$ and $0 = \sum_1^n h_i^{1/2} a\, h_i^{1/2}$. If Q is any conditional expectation onto $C^*(G^0)$, then

$$0 = Q(\sum_1^n h_i^{1/2} a h_i^{1/2}) = \sum_1^n h_i^{1/2} Q(a) h_i^{1/2} = \sum_1^n h_i Q(a)$$

$$= Q(\sum_1^n h_i a) = Q(a) .$$

Since Δ is closed and open in G, an arbitrary a in $C_c(G,\sigma)$ may be written $a = a_1 + a_2$ where a_1 is the restriction of a to Δ and supp a_2 does not meet Δ. Consequently, $Q(a) = a_1$. This shows that Q agrees with P on $C_c(G,\sigma)$, hence on $C^*_{red}(G,\sigma)$. To see that P is faithful, note that if $a \in C^*_{red}(G,\sigma)$, then $P(a^* * a)(u) = \int |a(x^{-1})|^2 \, d\lambda^u(x)$. Hence if $P(a^* * a) = 0$, then $a(x) = 0$ for all x.

Q.E.D.

<u>4.9. Definition :</u> Let A be a C^*-algebra and B an abelian sub C^*-algebra. We call <u>normalizer</u> of B (in A) the inverse semi-group

$\mathcal{N}(B) = \{a, \text{partial isometry of } A : d(a), r(a) \in B \text{ and } a(Bd(a))a^* = Br(a)\}$ where d(a) and r(a) denote the initial and final projections of a. An element $a \in \mathcal{N}(B)$ induces an isomorphism $s_a : b \to aba^*$ of Bd(a) onto Br(a) ; we also denote the corresponding partial homeomorphism of d(a) onto r(a) in the spectrum $\hat{B}$ of B by s_a. The inverse semi-group of partial homeomorphisms of $\hat{B}$ of the form s_a with $a \in \mathcal{N}(B)$ is called the ample <u>semi-group of (B,A)</u> (or of B when there is no ambiguity) and is denoted $\mathcal{G}(B)$.

The semi-group of partial isometries of B is denoted $\mathcal{U}(B)$. The unit spaces of $\mathcal{N}(B)$, $\mathcal{G}(B)$ and $\mathcal{U}(B)$ can all be identified with the Boolean algebra $\mathcal{B}$ of projections of B.

Remark : If B is a maximal abelian subalgebra of a C^*-algebra A,

$$\mathcal{B} \to \mathcal{U}(B) \to \mathcal{N}(B) \to \mathcal{G}(B) \to \mathcal{B}$$

is an exact sequence of inverse semi-groups (see 1.1.17). Indeed, if s_a is an idempotent in $\mathcal{G}(B)$, then $ab = ba$ for any $b \in B$, hence $a \in \mathcal{N}(B) \cap B = \mathcal{U}(B)$.

Recall that the ample semi-group of an r-discrete groupoid is the semi-group of its compact-open G-sets. In the case of a principal groupoid, G-sets are uniquely determined by the partial transformations they induce on the unit space. Therefore can be viewed as a semi-group of partial homeomorphisms of the unit space of the principal groupoid.

4.10. Proposition : Let G be an r-discrete principal groupoid with Haar system and let σ be a continuous 2-cocycle. Then the ample semi-group $\mathcal{G}(B)$ of the maximal abelian subalgebra $B = C^*(G^0)$ of the C^*-algebra $A = C^*_{red}(G,\sigma)$ coincides with the ample semi-group $\mathcal{G}$ of the groupoid.

roof : We first show that $\mathcal{G}$ is contained in $\mathcal{G}(B)$. If s is a compact-open G-set, ts characteristic function χ_s is a partial isometry in $C_c(G,\sigma)$ which normalizes B, s an obvious computation shows : $\chi_s * \chi_s^* = r(s)$, where r(s) is identified with its haracteristic function, $\chi_s^* * \chi_s = d(s)$, and $\chi_s * h * \chi_s^* = h^s$ for $h \in Bd(s)$, where $^s(u) = h(u\cdot s)$ if $u \in r(s)$ and 0 if $u \notin d(s)$. Hence χ_s induces the partial homeo- orphism $u \to u\cdot s$. Conversely, suppose that a is in $\mathcal{N}(B)$ and let $s = s_a$ the partial omeomorphism it induces on G^0. We want to show that its graph is a compact open G-set. in proposition 2.9 of [31, II], a simple computation shows that

$$a * h * a^*(u) = \int |a(y)|^2 \, h\circ d(y) d\lambda^u(y)$$

r $h \in C_c(d(s))$. By definition, this equals $h(u\cdot s)$ for $u \in r(s)$. We fix $u \in r(s)$. $y \in G^u_{d(s)}$ does not belong to s, then $d(y) \neq u\cdot s$ and there exists an non-negative nction $h \in C_c(d(s))$ such that $h\circ d(y) = 1$ and $h(u\cdot s) = 0$; this implies $a(y) = 0$. nce $a(y) = 0$ if $y \notin s$. Moreover, $|a(y)| = 1$ if y belongs to s. Since a is in (G), s must be a compact open set of G.

Q.E.D.

A last property of the pair $(C^*_{red}(G,\sigma), C^*(G^0))$ needs to be interpreted in terms of the groupoid. This is the notion of regular abelian subalgebra introduced in Dixmier [17] in the context of von Neumann algebras.

4.11. Definition : An abelian sub-$*$-algebra B of a C^*-algebra A is said to be regular if the linear span of the elements of the form ab, where $a \in \mathcal{N}(B)$ and $b \in B$, is dense in A.

4.12. Proposition : Let G be an r-discrete groupoid with Haar system and σ a continuous 2-cocycle. Then $C^*(G^0)$ is a regular subalgebra of $C^*_{red}(G,\sigma)$ iff G can be covered with compact open G-sets.

Proof : If G can be covered with compact G-sets, then one can, by using a partition of the unity, write any $f \in C_c(G,\sigma)$ as a sum of functions supported on compact open G-sets and a function supported on the compact open G-set s may be written under the form $\chi_s * h$ where $h \in C_c(G^0)$. Conversely, if the space of continuous functions supported on compact open G-sets is dense in $C^*_{red}(G,\sigma)$, they cannot all vanish at a given point x of G. Consequently such point x is contained in a compact open G-set.

Q.E.D.

The properties of the subalgebra $C^*(G^0)$, when G is an r-discrete principal groupoid, may be summarized by introducing, as in definition 3.1 of [31, II], the notion of Cartan subalgebra. Recall that the ample semi-group of an r-discrete principal groupoid with Haar system has the property of acting relatively freely on the unit space (1.2.14), in the sense that the set of fixed points of each of its elements is open.

4.13. Definition : An abelian sub-$*$-algebra B of a C^*-algebra will be called a Cartan subalgebra if it has the following properties :

(i) it is maximal abelian ;

(ii) it is regular ;

(iii) its ample semi-group $\mathcal{G}(B)$ acts relatively freely on its spectrum $\hat{B}$; and

(iv) the exact sequence $\mathcal{B} \to \mathcal{U}(B) \to \mathcal{N}(B) \overset{s}{\to} \mathcal{G}(B) \to \mathcal{B}$ splits in the sense that there exists a section k for s satisfying $k(se) = k(s)e$, $k(es) = ek(s)$ and $k(e) = e$, for every e in $\mathcal{B}$ and s in $\mathcal{G}(B)$.

Question : Is (iv) independent of (i) - (iii) ?

4.14. Proposition : Let G be an r-discrete principal groupoid admitting a cover of compact open G-sets and let σ be a continuous 2-cocycle. Then $C^*(G^0)$is a Cartan subalgebra of $C^*_{red}(G,\sigma)$.

Proof : It is maximal abelian by 4.7.(ii), regular by 4.12 and its ample semi-group $\mathcal{G}$, which is the ample semi-group of G by 4.10, acts relatively freely on G^0 by 1.2.13. A section for $\mathcal{N}(C^*(G^0)) \to \mathcal{G}$ is given by $k(s) = \chi_s$ where χ_s is the characteristic function of the G-set s.

Q.E.D.

This proposition admits a converse.

4.15. Theorem : Let B be a Cartan subalgebra of a separable C^*-algebra A.

(i) There exists an r-discrete principal groupoid G admitting a cover by compact open G-sets, a continuous 2-cocycle σ and a $*$-homomorphism ϕ of $C^*(G,\sigma)$ onto A which carries faithfully $C^*(G^0)$ onto B and the ample semi-group of G onto the ample semi-group of B.

(ii) The groupoid G is unique up to isomorphism and the 2-cocycle σ is unique up to a coboundary.

(iii) If G is amenable, the $*$-homomorphism ϕ is an isomorphism and B is the image of a unique conditional expectation, which is faithful.

Proof :

(i) The ample semi-group $\mathcal{G}(B)$ of B is an inverse semi-group of partial homeomorphisms of $\hat{B}$, defined on compact open sets. By 4.13. (iii) and 1.2.13, the principal groupoid G associated to it has a structure of r-discrete groupoid with Haar system such that $\hat{B}$ becomes its unit space and $\mathcal{G}(B)$ its ample semi-group. Let k be a section for s as in 4.13 (iv). By definition of s, it satisfies the covariance property $k(t)$ $a\, k(t)^* = a^t$ for each $t \in \mathcal{G}(B)$ and each $a \in \mathcal{U}(B)$. Hence, the extension is compatible with the action of $\mathcal{G}(B)$ on $\mathcal{U}(B)$ (see 1.1.17). As in 1.1.17, this extension is defined by a 2-cocycle $\tilde{\sigma} \in Z^2(\mathcal{G}(B),\mathcal{U}(B))$. As in 1.2.14, there exists a unique continuous

2-cocycle $\sigma \in Z^2(G,\mathbb{T})$ such that

$$\tilde{\sigma}(s,t)(u) = \sigma(us,ust)$$

for every $s,t \in \mathcal{G}(B)$. We try to define a map ϕ of $C_c(G,\sigma)$ into A by the formula $\phi(\sum_1^n h_i x_{s_i}) = \sum_1^n h_i k(s_i)$ where $h_i \in C_c(G^0) \subset C^*(G^0) = B$ and $s_i \in \mathcal{G}(B)$. The map ϕ is well defined. First, any element of $C_c(G,\sigma)$ may be written $\sum_1^n h_i x_{s_i}$. Second, suppose that $\sum_1^n h_i x_{s_i} = 0$: find disjoint compact open G-sets t_j, $j = 1,\dots,m$ such that each s_i may be expressed as an union of t_j's. We may write $s_i = \bigcup_j \varepsilon_{ij} t_j$ where $\varepsilon_{ij} = 0$ or 1 and $0t_j = \emptyset$, $1t_j = t_j$. Then $k(s_i) = \sum_j \varepsilon_{ij} k(t_j)$ and $x_{s_i} = \sum_j \varepsilon_{ij} x_{t_j}$. The equality $\sum_j(\sum_i \varepsilon_{ij} h_i) x_{t_j} = \sum_i h_i x_{s_i} = 0$ implies $\sum_i \varepsilon_{ij} h_i|_{r(t_j)} = 0$ for each j, because the t_j's are disjoint. Therefore $\sum_i h_i k(s_i) = \sum_j(\sum_i \varepsilon_{ij} h_i) k(t_j) = 0$. The same argument shows that ϕ is one-to-one. We note that ϕ is a $*$-homomorphism. For if s is a G-set, then

$$x_s^* = \tilde{\sigma}(s^{-1},s)^* x_{s^{-1}},$$

$$\phi(x_s^*) = \tilde{\sigma}(s^{-1},s)^* k(s^{-1})$$

$$= [k(s^{-1})\, k(s)\, k(s^{-1}s)^*]^*\, k(s^{-1})$$

$$= k(s)^* = \phi(x_s)^*;$$

and if s and t are G-sets, then

$$x_s * x_t = \tilde{\sigma}(s,t)\, x_{st}$$

$$\phi(x_s * x_t) = \tilde{\sigma}(s,t)\, k(st)$$

$$= k(s)\, k(t)\, k(st)^*\, k(st) = k(s)\, k(t)$$

$$= \phi(x_s)\, \phi(x_t).$$

The map ϕ is continuous when $C_c(G,\sigma)$ has the inductive limit topology. Indeed, let (f_i) converge to f in the inductive limit topology ; multiplying by a finite partition of unity, we may assume that the f_i's and f have their support contained in a common compact open G-sets ; the assertion is now obvious. We may apply theorem 1.21 (or rather, its corollary 1.22) to conclude that ϕ is continuous for the C^*-norm of $C_c(G,\sigma)$. Since A is separable ; B is separable and $\mathcal{G}(B)$ is countable, hence G is second countable. We have already noted (1.3.28) that a second countable r-discrete groupoid has sufficiently many non-singular G-sets. Thus ϕ extends to a $*$-homomorphism of $C^*(G,\sigma)$ into A. It is onto because its range contains the elements ab with

$a \in \mathcal{G}(B)$ and $b \in B$ and B is regular .

(ii) The groupoid G is uniquely defined by B and its ample semi-group. The 2-cocycle σ is determined up to a coboundary by the extension

$$\mathcal{B} \to \mathcal{U}(B) \to \mathcal{N}(B) \to \mathcal{G}(B) \to \mathcal{B} .$$

(iii) If G is amenable, proposition 3.2 shows that $C^*(G,\sigma) = C^*_{red}(G,\sigma)$. The kernel of ϕ is an ideal of $C^*_{red}(G,\sigma)$ which intersects $C_c(G,\sigma)$ trivially, hence it is trivial by 4.6. The last assertion results from 4.8.

Q.E.D.

4.16. Remark : It will be given in 3.2.5 an example of a C^*-algebra with two maximal abelian sub C^*-algebras, one of which is a Cartan subalgebra, the other satisfies (i)(ii) but not (iii) of 4.13 and is the image of a unique faithful conditional expectation.

Let us conclude this section by recalling some facts, due to P. Hahn, pertinent to the regular representations of a principal groupoid. The case of an r-discrete principal groupoid is studied in [31].

4.17. Proposition : (P.Hahn [45]). Let G be a second countable locally compact groupoid with Haar system, σ a continuous 2-cocycle and μ a quasi-invariant measure on G^0. Then

(i) the σ-regular representation on μ (defined in 1.8) is a factor representation iff μ is ergodic,

(ii) it is of type I (resp. II_1, II_∞, III) iff μ is of type I (resp. II_1, II_∞, III) (defined in 1.3.13).

Proof : The assertion (i) and part of the assertion (ii) result from his theorem 5.1. The rest results from his theorems 5.4 and 5.5.

5. Automorphisms Groups, KMS States and Crossed Products

This section illustrates the use of groupoids in the study of basic problems for C^*-algebras. As usual G denotes a locally compact groupoid with Haar system $\{\lambda^u\}$ and σ a continuous 2-cocycle. Here A denotes a locally compact abelian group with dual group $\hat{A}$. The value of the character $\xi \in \hat{A}$ at $a \in A$ is written (ξ,a). Let $c \in Z^1(G,A)$ be a continuous one-cocycle. Define, for each $\xi \in \hat{A}$,

$$\alpha_\xi(f)(x) = (\xi,c(x))f(x) \quad \text{for } f \in C_c(G,\sigma).$$

5.1. Proposition : Let $c \in Z^1(G,A)$ and α be as above. Then

(i) α_ξ is an automorphism of $C_c(G,\sigma)$;

(ii) α_ξ extends to an automorphism of $C^*(G,\sigma)$;

(iii) $(C^*(G,\sigma),\hat{A},\alpha)$ is a C^*-dynamical system (see 7.4.1 in [60]), in other words α is a continuous homomorphism of $\hat{A}$ into the group $\mathrm{Aut}(C^*(G,\sigma))$ of automorphisms of $C^*(G,\sigma)$ equipped with the topology of pointwise convergence ; and

(iv) α leaves $C^*(G^0)$ pointwise fixed.

Proof :

(i) This is a routine verification.

(ii) First, one notes that α_ξ is isometric with respect to the $\| \|_I$ norm :

$$\int |\alpha_\xi(f)|(x)d\lambda^u(x) = \int |f(x)|d\lambda^u(x).$$

Hence α_ξ is continuous with respect to the C^*-norm and so is $\alpha_\xi^{-1} = \alpha_{\xi^{-1}}$

(iii) It is clear that $\alpha : \hat{A} \to \mathrm{Aut}(C^*(G,\sigma))$ is a group homomorphism. Let us check its continuity. It suffices to check that the map $\xi \to \alpha_\xi f$ is continuous for any $f \in C_c(G,\sigma)$ and for the topology of the $\| \|_I$ norm. Let K be the support of f. For any $\varepsilon > 0$, there exists a neighborhood V of ξ in $\hat{A}$ such that for $\eta \in V$ $|(\eta,c(x)) - (\xi,c(x))| \leq \varepsilon$ for any $x \in K$. Then $\|\alpha_\eta f - \alpha_\xi f\|_{I,r} = \sup_u \int |\alpha_\eta f - \alpha_\xi f| d\lambda^u \leq \varepsilon \|f\|_{I,r}$.

Hence, $\|\alpha_\eta f - \alpha_\xi f\|_I \leq \varepsilon \|f\|_I$.

(iv) Clear.

Q.E.D.

In the case when A is the group of real numbers, one may define a linear map δ on the domain $D(\delta) = C_c(G,\sigma)$ by

$$(\delta f)(x) = ic(x)f(x)$$

5.2. Proposition. Let $c \in Z^1(G,\mathbb{R})$, and let α and δ be as above. Then

(i) $C_c(G,\sigma)$ consists of entire analytic elements for α ; and

(ii) δ is a $*$-derivation and a pregenerator for α. (A reference for unbounded derivations on C^*-algebras is [65]).

Proof : We note that for any continuous function ϕ on G and $f \in C_c(G,\sigma)$,

$$\|\phi f\| \leq \|\phi f\|_I \leq (\sup_{x\in K} |\phi(x)|)\ \|f\|_I$$

where ϕf denotes the pointwise product and K is the support of f. Therefore, $C_c(G,\sigma)$ is in the domain of the generator of α and δ is its restriction, and

$$\left\|\frac{\alpha_t f - f}{t} - \delta f\right\| \leq \sup_K \left|\frac{e^{itc(x)} - 1}{t} - ic(x)\right| \ \|f\|_I .$$

The same argument shows that $\delta^n f$ exists for any integer n and

$$\|\delta^n f\| \leq (\sup_K |c(x)|)^n \ \|f\|_I .$$

This proves the first assertion. Also, δ is closable and its closure generates an automorphism group which can be nothing but α. Hence, the closure of δ is the generator of α.

Q.E.D.

Let us say that an automorphism group α_ξ of a C^*-algebra is inner if there exists a group of unitaries U_ξ in the multiplier algebra such that

(i) $\alpha_\xi(A) = U_\xi A\, U_\xi^*$ for any element A of the C^*-algebra, and

(ii) $\xi \to U_\xi$ is continuous for the strict topology. (Recall that the strict topology on the multiplier algebra is defined by the semi-norms $A \to \|AB\|$ and $A \to \|BA\|$ for B in the original algebra).

5.3. Proposition : Let $c \in Z^1(G,A)$ and let α be the associated automorphism group of $C^*(G,\sigma)$.

(i) If $c \in B^1(G,A)$, then α is inner.

(ii) If G is r-discrete, principal and amenable, the converse holds.

Proof :

(i) One first observes that any bounded continuous function on G^0 defines in the obvious way an element of the multiplier algebra of $C^*(G,\sigma)$. If $c(x) = b\circ r(x) - b\circ d(x)$ where b is a continuous function on G^0, then for each ξ in $\hat{A}$, $U_\xi(u) = (\xi, b(u))$ defines a unitary element of the multiplier algebra and for $f \in C_c(G,\sigma)$, $\alpha_\xi(f)(x) = (\xi, c(x))f(x) = (U_\xi f\, U_\xi^*)(x)$. The continuity of $\xi \to U_\xi$ is checked as in 5.1 (iii).

(ii) If $\alpha_\xi(A) = U_\xi A\, U_\xi^*$, then U_ξ commutes with every element of $C^*(G^0)$ and hence is itself diagonal (see section 4 - we have not considered the multiplier algebra there, but its elements can also be viewed as continuous functions on G). Therefore U_ξ is of the form $U_\xi(u) = (\xi, b(u))$ where b is a continuous function on G^0 and $c(x) = b\circ r(x) - b\circ d(x)$.

Q.E.D.

As an example, let us interpret the theorem 4.8 of the first chapter. We assume that G is an r-discrete, principal and amenable groupoid with compact unit space. By 4.6 of this chapter, $C^*(G,\sigma)$ is simple iff G is minimal. Let $c \in Z^1(G,\mathbf{R})$ and assume that c is bounded. This amounts to saying the associated derivation δ is bounded, or equivalently, that the associated automorphism group is norm continuous. Then the range of c $R(c)$ is compact and the asymptotic range $R_\infty(c)$ is zero. The theorem states that if G is minimal and c bounded, then c is in $B^1(G,A)$. In other words, if $C^*(G,\sigma)$ is simple and δ bounded, then δ is inner. This is a particular case of a well known result of Sakai ([64], 4.1.11).

When G is r-discrete, principal and amenable, the asymptotic range $R_\infty(c)$ of a cocycle $c \in Z^1(G,A)$ can be identified as the Connes spectrum (see [60], 8.8.2) of the associated automorphism group. This is heuristically clear when one compares both definitions :

$$R_\infty(c) = \cap R(c_U)$$

where U runs over all non-empty open sets in G^0, c_U is the restriction of c to $G_{|U}$, and $R(c)$ is the closure of $c(G)$; while

$$\Gamma(\alpha) = \cap \mathrm{Sp}(\alpha|B)$$

where B runs over all α-invariant, hereditary non-zero sub C^*-algebras of $C^*(G,\sigma)$, and $\mathrm{Sp}(\alpha)$ is the Arveson spectrum of α([60] 8.1.6.). It can be seen in our case that

$Sp(\alpha) = R(c)$ and that for every open set U in G^0, $C^*(G_{|U},\sigma)$ may be viewed as an α-invariant hereditary subalgebra of $C^*(G,\sigma)$. (This may be done in a fashion analogous to 4.4). However, in order to avoid the explicit determination of the hereditary subalgebras of $C^*(G,\sigma)$, we will use an alternate definition of $\Gamma(\alpha)$ ([60], 8.11.8) which uses only the ideals of the cross-product algebra $C^*(G,\sigma) \times_\alpha \hat{A}$. This will be done in 5.8.

Given a cocycle c in $Z^1(G,\mathbb{R})$ and $\beta \in [0, +\infty]$, the (c,β)KMS condition for a measure μ on G^0 has been defined in 1.3.15. It is time to justify this terminology. We have seen how a one-parameter automorphism group α of $C^*(G,\sigma)$ is associated to c. On the other hand, composing μ with the restriction map from $C_c(G,\sigma)$ onto $C_c(G^0)$, one obtains a positive linear functional $\phi = \phi_\mu$ on $C_c(G,\sigma)$. A positive linear functional on $C_c(G,\sigma)$ continuous for the inductive limit topology - an equivalent term is "positive type measure" - will be called here a weight on $C^*(G,\sigma)$. This does not agree with the usual definition of a weight on a C^*-algebra (see [12], page 61), because $C_c(G,\sigma)$ is not always a hereditary subalgebra of $C^*(G,\sigma)$, but it is convenient here. If G is r-discrete and μ a probability measure, ϕ is a state. We note that, with above notations, ϕ is α-invariant since c vanishes on G^0.

5.4. Proposition : Let $c \in Z^1(G,\mathbb{R})$, $\beta \in [0, +\infty]$ and μ be a measure on G^0. The automorphism group associated with c is denoted by α and the weight associated with μ is denoted by ϕ. Then the following properties are equivalent :

(i) the weight ϕ satisfies the (α,β)KMS condition (see [60] 8.12.2 or [65] 6.1) ; and

(ii) the measure μ satisfies the (c,β)KMS condition (1.3.15). Moreover, if G is principal and β finite, any weight ϕ which satisfies the (α,β)KMS condition arises from a measure μ on G^0.

Proof : We first consider the case when β is finite. Replacing c by βc, we may assume that $\beta = 1$.

(i) ==> (ii) Since ϕ is 1 - KMS for α, we find that for any $f,g \in C_c(G,\sigma)$ and any $t \in \mathbb{R}$

$$\phi[\alpha_t(f) * g] = \phi[g * \alpha_{t+i}(f)],$$

because f is analytic for α(5.2.(i)). Let us evaluate both expressions.

For the first,

$$\phi[\alpha_t(f) * g] = \int e^{itc(y)} f(y)\, g(y^{-1}) d\nu(y), \text{ where } \nu = \int \lambda^u d\mu(u),$$

while for the second,

$$\begin{aligned}\phi[g * \alpha_{t+i}(f)] &= \int g(y)\, e^{i(t+i)c(y^{-1})} f(y^{-1})\, d\nu(y) \\ &= \int e^{itc(y)} f(y)\, g(y^{-1})\, e^{-c(y)}\, d\nu^{-1}(y).\end{aligned}$$

In particular, for any $f \in C_c(G)$

$$\int f(y)\, d\nu(y) = \int f(y)\, e^{-c(y)}\, d\nu^{-1}(y),$$

so that $D = \dfrac{d\nu}{d\nu^{-1}}$ exists and is equal to e^{-c} (ν a.e.).

(ii) ⟹ (i) The same computation shows that, if μ is quasi-invariant with $\dfrac{d\nu}{d\nu^{-1}} = e^{-c}$, then

$$\phi[\alpha_t(f) * g] = \phi[g * \alpha_{t+i}(f)] \quad \text{for any } f, g \in C_c(G,\sigma).$$

Second, we consider the case when β is infinite. The ∞ - KMS condition asserts that for any $f \in C_c(G,\sigma)$,

$$- i\phi(f^* * \delta(f)) \geq 0.$$

After a computation, this becomes

$$\int |f|^2 c\, d\nu^{-1} \geq 0, \text{ where } \nu = \int \lambda^u d\mu(u).$$

Hence, ϕ satisfies the ∞ - KMS condition iff c is non-negative on the support of ν^{-1}, which is the inverse image under d of the support of μ. But this is just the ∞ - KMS condition for μ, namely supp $\mu \subset \text{Min}(c) = \{u \in G^0 : c_{|G_u} \geq 0\}$.

Finally suppose that G is principal, β finite and that the weight ϕ, corresponding to the positive type measure μ, satisfies the (α,β) KMS condition. Then for any $f,g \in C_c(G,\sigma)$ and any $t \in \mathbb{R}$, we have

$$\phi[\alpha_t(f) * g] = \phi[g * \alpha_{t+i}(f)].$$

Using a left approximate identity for $C_c(G,\sigma)$ endowed with the inductive limit topology, one gets $\phi(hg) = \phi(gh)$ for $g \in C_c(G,\sigma)$ and $h \in C_c(G^0)$. We want to show that the support of μ is contained in G^0. Suppose that $g \in C_c(G)$ and supp $g \cap G^0 = \emptyset$. Since G is principal, supp g may be covered by open sets U such that $d(U) \cap r(U) = \emptyset$. Using a partition of the unity, we may write $g = \sum_{1}^{n} g_i$ with $d(\mathrm{supp}\ g_i) \cap r(\mathrm{supp} g_i) = \emptyset$. If we choose $h \in C_c(G^0)$ which takes the value 1 on $d(\mathrm{supp}\ g_i)$ and 0 on $r(\mathrm{supp}\ g_i)$, we have $\phi(g_i) = \phi(hg_i) = \phi(g_i h) = 0$, hence $\phi(g) = 0$.

Q.E.D.

5.5. Remarks :

a. Since it is important to determine all KMS weights of a group of automorphisms, we give the following complement for $\beta = \infty$. Let ϕ be a weight corresponding to a positive type measure μ on G. Then one can show that ϕ satisfies the (α,β) KMS condition only if $\mathrm{supp}\mu \subset c^{-1}(0) \cap d^{-1}(\mathrm{Min}(c))$. In particular, if Min(c) is reduced to one element u, then $c^{-1}(0) \cap d^{-1}(\mathrm{Min}(c))$ is also reduced to $\{u\}$ because Min(c) is $c^{-1}(0)$ invariant (1.3.16 (iv)). Thus there is only one KMS weight at ∞, namely, the point mass at μ.

b. Given an (α,β) KMS weight ϕ, it is natural to look at the GNS representation L it generates. It is the representation induced by μ in the sense of 2.7. It acts on $L^2(G,\nu^{-1})$ by left convolution. Let $\mathfrak{L}$ be the von Neumann algebra it generates. There exists a unique normal semi-finite weight $\tilde{\phi}$ on $\mathfrak{L}$ which extends ϕ in the sense that $\phi(f) = \tilde{\phi}\circ L(f)$ for $f \in C_c(G,\sigma)$, and there is a unique automorphism group $\tilde{\alpha}$ which extends α,

$$\tilde{\alpha}_t \circ L(f) = L\circ\alpha_t(f) \text{ for } f \in C_c(G,\sigma).$$

Let H be the operator of multiplication by c on $L^2(G,\nu^{-1})$. Then $\tilde{\alpha}$ is given by

$$\tilde{\alpha}_t(A) = e^{itH} A e^{-itH} .$$

The operator H is interpreted as the energy operator in this representation.

Let us consider the case β finite. We first assume $\beta = 1$. The representation L is in standard form. It is the regular representation on μ and appears as the left representation of the generalized Hilbert algebra introduced in 1.10. In particular $\tilde{\alpha}$ is the modular group of the faithful normal semi-finite weight $\tilde{\phi}$. The relation between the modular operator Δ, which is given by multiplication by the R-N derivative D, and the energy operator H is $\Delta = e^{-H}$. In the case when β is arbitrary but finite, we replace c by βc and obtain the relation $\Delta = e^{-\beta H}$ between Δ and H.

In the case $\beta = \infty$, the representation L is no longer in standard form. The

∞-KMS condition is precisely the requirement $H \geq 0$. One says that a vector $\xi \in L^2(G,\nu^{-1})$ has zero energy if $\xi(x) = 0$ for ν^{-1} a.e. x such that $c(x) = 0$ and that ϕ is a physical ground weight ([65], definition 5.2) if the space of vectors of zero energy is one-dimensional. A necessary condition is that μ is a point mass. In fact, the point mass at $u \in Min(c)$ defines a physical ground weight iff $[u] \cap Min(c) = \{u\}$, where $[u]$ is the orbit of u.

c. Suppose that $c \in B^1(G,\mathbb{R})$, $c(x) = b\circ r(x) - b\circ d(x)$. Then we know that α is inner. It is implemented by the group of unitaries $U_t(u) = e^{itb(u)}$. If we think of b as the energy function, the interpretation of Min(c) is clear : $Min(c) = \{u \in G^0$: the restriction of b to $[u]$ reaches its minimum at u$\}$. In the general case, we will call c the <u>energy cocycle</u> of the system.

Given a cocycle $c \in Z^1(G,A)$, we have defined the C^*-dynamical system $(C^*(G,\sigma),\hat{A},\alpha)$. Our last task in this section is to identify the crossed product C^*-algebra $C^*(G,\sigma) \times_\alpha \hat{A}$ as the C^*-algebra of the skew product, that is, $C\ (G^*(c),\sigma)$. Let us recall some notations and introduce new ones : G is a locally compact groupoid with Haar system (λ^u) ; σ is a continuous 2-cocycle in $Z^2(G,\mathbb{T})$; A is a locally compact abelian group, noted multiplicatively ; its dual group $\Gamma = \hat{A}$ will be noted multiplicatively too ; and c is a continuous 1-cocycle in $Z^1(G,A)$. The skew product G(c) is the locally compact groupoid obtained by defining on G x A the multiplication $(x,a)(y,ac(x)) = (xy,a)$ and the inverse $(x,a)^{-1} = (x^{-1},ac(x))$. A composable pair will be written (x,y,a) instead of $((x,a),(y,ac(x)))$. The groupoid G(c) has the Haar system $(\lambda^{u,a} = \lambda^u \times \delta_a)$. A cocycle on G lifts to a cocycle on G(c), for example, we define $\sigma(x,y,a) = \sigma(x,y)$.

Let (E,Γ,α) be a Banach $*$-algebra dynamical system, that is, E is a Banach $*$-algebra, Γ a locally compact group and α a continuous homomorphism of Γ into Aut(E) equipped with the topology of pointwise convergence. Recall that $L^1(\Gamma,E)$ is the space of E-valued functions on Γ integrable with respect to the Haar measure of Γ (in our case, Γ is abelian). It is made into a Banach $*$-algebra with the operations :

$$f * g(\xi) = \int f(\eta)\ \alpha_\eta[g(\eta^{-1}\xi)]d\eta,$$

$$f^*(\xi) = \overline{f(\xi^{-1})},$$

and the norm $\|f\|_1 = \int \|f(\xi)\| d\xi$. A covariant representation of the system on a Hilbert space $\mathcal{H}$ consists of a continuous unitary representation V of Γ on $\mathcal{H}$ and a norm - decreasing nondegenerate representation M of E such that $V(\xi)\, M(e)\, V(\xi)^* = M[\alpha_\xi(e)]$. A covariant representation (V,M) has an integrated form. Namely,

$$L(f) = \int M[f(\xi)] \quad V(\xi)d\xi$$

defines a non-degenerate representation of L on $\mathcal{H}$. Conversely, if E has a bounded approximate identity, any non-degenerate representation of $L^1(\Gamma,E)$ is an integrated form and the correspondence is bijective. All this is well known and we refer to [20] for further details. If E is a C^*-algebra, the crossed product C^*-algebra $E \times_\alpha \Gamma$ is the enveloping C^*-algebra of $L^1(\Gamma,E)$.

Recall that we defined the norm $\| \ \|_I$ on $C_c(G,\sigma)$ by

$$\|f\|_I = \max \{\sup_u \int |f| d\lambda^u, \ \sup_u \int |f| d\lambda_u\}$$

It is a $*$-algebra norm on $C_c(G,\sigma)$. We denote the completion of $C_c(G,\sigma)$ in the norm $\| \ \|_I$ by $L^I(G,\sigma)$. One annoying problem with this Banach $*$-algebra is the existence of a bounded approximate identity. It can be established without difficulty in the r-discrete case (take a bounded approximate identity for $C^*(G^0)$) and when G is a transformation group (take the pointwise product $h_i e_i$, where e_i is the characteristic function of a symmetric neighborhood of the identity of the group, normalized for the left Haar measure, and e_i a bounded approximate identity for $C^*(G^0)$), but I don't know if it always exists in the general case. Note that, as a Banach space, $L^I(G(c),\sigma)$ is $C_0(A,L^1(G,\sigma))$, the space of $L^I(G,\sigma)$-valued continuous functions on A which vanish at infinity.

<u>5.6. Lemma :</u> Let E be a separable Banach space, Γ a locally compact abelian group and α a continuous homomorphism of Γ into the group of isometries of E equipped with the topology of pointwise convergence. Then

$$f \mapsto \hat{f}(a) = \int_\Gamma \alpha_\xi^{-1}[f(\xi)] \quad \overline{(\xi,a)}\ d\xi$$

defines a norm-decreasing linear map with dense range from $L^1(\Gamma,E)$ into $C_0(\hat{\Gamma},E)$.

<u>Proof :</u> Clearly, $\hat{f}(a)$ is well defined and $\|\hat{f}(a)\| \le \|f\|$. By the Lebesgue dominated convergence, $\hat{f}$ is a continuous function from $\hat{\Gamma}$ to E. If $\hat{f}$ is a decomposable element of $L^1(\Gamma,E)$, that is, an element of the form $f(\xi) = \sum_1^n f_i(\xi)\, e_i$, where $f_i \in L^1(\Gamma)$ and $e_i \in E$ for $i = 1,\dots,n$, then $\hat{f}$ vanishes at infinity. Since decomposable elements are dense in $L^1(\Gamma,E)$, the map sends $L^1(\Gamma,E)$ into $C_0(\hat{\Gamma},E)$. We want to show that it has

dense range. Note that the map β, defined by $\beta(f)(\xi) = \alpha_\xi^{-1}[f(\xi)]$ is an isometry of $L^1(\Gamma,E)$. Therefore, it suffices to consider the map $f \to \hat{f}$, where $\hat{f}(a) = \int f(\xi)\,\overline{(\xi,a)}d\xi$. Since the Fourier transform from $L^1(\Gamma)$ into $C_0(\hat{\Gamma})$ has dense range, every decomposable element of $C_0(\hat{\Gamma},E)$ lies in the range closure of the map $f \to \hat{f}$. Since decomposable elements are dense in $C_0(\hat{\Gamma},E)$, the range closure is $C_0(\hat{\Gamma},E)$.

Q.E.D.

5.7. Theorem : Let G, σ, A, c and α be as above. Assume that $L^I(G,\sigma)$ has a bounded approximate unit. Then the crossed-product C^*-algebra $C^*(G,\sigma) \times_\alpha \hat{A}$ is isomorphic to the C^*-algebra $C^*(G(c),\sigma)$ of the skewproduct.

Proof : Since the automorphism α_ξ of $C_c(G,\sigma)$, given by $\alpha_\xi(f)(x) = (\xi,c(x))f(x)$, preserves the $\| \; \|_I$ norm, it extends to an automorphism of $L^I(G,\sigma)$. The continuity of $\alpha : \Gamma \to \mathrm{Aut}(L^I(G,\sigma))$ is established as in 5.1.(iii). By 5.6, the map from $L^1(\hat{A},L^I(G,\sigma))$ to $C_0(A,L^I(G,\sigma)) = L^I(G(c),\sigma)$ defined by $f \to \hat{f}(a) = \int_\Gamma \alpha_\xi^{-1}[f(\xi)]\,\overline{(\xi,a)}d\xi$ is norm-decreasing and has dense range. It is a straightforward computation to check that it is a $*$-algebra homomorphism. Let us just write down the relevant formulas

$$\hat{f}(x,a) = \int f(x,\xi)\overline{(\xi,ac(x))}\,d\xi$$

for $f,g \in C_c(G \times \hat{A}) \subset L^1(\hat{A},L^I(G,\sigma))$,

$$f * g\,(x,\xi) = \iint f(y,\eta)\,g(y^{-1}x,\eta^{-1}\xi)\,(\eta,c(y^{-1}x))\sigma(y,y^{-1}x)d\lambda^{r(x)}(y)d\eta$$

$$f^*(x,\xi) = \bar{f}(x^{-1},\xi^{-1})\,\bar{\sigma}(x,x^{-1})\,(\xi,c(x))$$

and for $f,g \in C_0(A,C_c(G)) \subset L^I(G(c),\sigma)$,

$$f * g\,(x,a) = \int f(y,a)\,g(y^{-1}x,ac(y))\,\sigma(y,y^{-1}x)\,d\lambda^{r(x)}(y)$$

$$f^*(x,a) = \bar{f}(x^{-1},ac(x))\,\bar{\sigma}(x,x^{-1}).$$

Composing with the homomorphism of $L^I(G(c),\sigma)$ into $C^*(G(c),\sigma)$, we obtain a (norm-decreasing) $*$-algebra homomorphism π from $L^1(A,L^I(G,\sigma))$ into $C^*(G(c),\sigma)$ which has dense range. If L is a (non-degenerate) representation of $C^*(G(c),\sigma)$, $L\circ\pi$ is a non-degenerate representation of $L^1(\hat{A},L^I(G,\sigma))$. There exists a covariant representation (V,M) of $(\hat{A}, L^I(G,\sigma)$, of which $L\circ\pi$ is the integrated form. By definition of $C^*(G,\sigma)$, M decreases its C^*-norm and we obtain the estimate

$$\|L\circ\pi(f)\| = \|\int M|f(\xi)|V(\xi)d\xi\|$$
$$\leq \int \|f(\xi)\|d\xi = \|f\|_1$$

where $\| \ \|_1$ is the norm of $L^1(\hat{A},C^*(G,\sigma))$. Therefore, $L\circ\pi$ extends to a representation of $L^1(\hat{A},C^*(G,\sigma))$ and ipso facto to a representation of its enveloping C^*-algebra $C^*(G,\sigma)\times_\alpha \hat{A}$. We have $\|L\circ\pi(f)\| \le \|f\|$, where $\| \ \|$ is the norm of $C^*(G,\sigma)\times_\alpha A$. We conclude that $\|\pi(f)\| \le \|f\|$ and that π extends to a $*$-homomorphism from $C^*(G,\sigma)\times_\alpha \hat{A}$ to $C^*(G(c),\sigma)$. It is onto, because its range is dense and closed. Let us show that it is one-to-one, or, equivalently, isometric. Let L be the representation of $C^*(G,\sigma)\times_\alpha \hat{A}$ induced by the representation M of $C^*(G,\sigma)$. We will assume that M is the integrated form (cf. theorem 1.20) of the σ-representation $(\mu,\mathcal{K},M)$ of G. Let $H = \Gamma(\mathcal{K})$ (the space of square integrable sections of $\mathcal{K}$) be its representation space. By definition, L acts on $L^2(\hat{A};H)$ by

$$L(f)\phi(\gamma) = \int_\Gamma M[\alpha_\gamma^{-1}(f(\xi))] \quad \phi(\xi^{-1}\gamma)d\xi$$

where $f \in L^1(\hat{A};C^*(G,\sigma))$ and $\phi \in L^2(\hat{A};H)$. Let us consider the following σ-representation $(\mu\times\lambda,\mathcal{H},\hat{L})$ of $G(c)$: λ is the Haar measure of A (we have observed in 3.8 that $\mu\times\lambda$ is quasi-invariant), $\mathcal{H}_{u,a} = \mathcal{K}_u$; and $\hat{L}(x,a) : \mathcal{H}_{(d(x),ac(x))} \to \mathcal{H}_{(r(x),a)}$ is given by $\hat{L}(x,a) = M(x)$. Its integrated form acts on $\Gamma(\mathcal{H})$ by

$$\hat{L}(\hat{f})\phi(u,a) = \int \hat{f}(x,a)\ \hat{L}(x,a)\phi(d(x),ac(x))\ D^{-1/2}(x)d\lambda^u(x),$$

for $\hat{f} \in C_c(G(c),\sigma)$, $\phi \in \Gamma(\mathcal{H})$, where D is the modular function of μ. We may identify $\Gamma(\mathcal{H})$ with $L^2(A,H)$ in an obvious fashion, where $H = \Gamma(\mathcal{K})$, and we may define the Fourier transform $\mathcal{F}$ from $L^2(A,H)$ to $L^2(\hat{A};H)$ by $\mathcal{F}\phi(a) = \int \phi(\gamma)\overline{(\gamma,a)}\ d\gamma$. Of course, $\mathcal{F}$ is an isometry. It is then a straightforward computation to check that

$$\mathcal{F}\circ L(f) = \hat{L}(\hat{f}) \circ \mathcal{F}$$

for any $f \in C_c(\hat{A}\times G)$, where $\hat{f} = \pi(f)$. The relevant formulas are

$$L(f)\phi(u,\gamma) = \iint \overline{(\gamma,c(x))}\ f(x,\xi)\ M(x)\phi(d(x),\xi^{-1}\gamma)\ D^{-1/2}(x)\ d\lambda^u(x)d\xi,$$

$$\hat{L}(\hat{f})\phi(u,a) = \int \hat{f}(x,a)\ M(x)\phi(d(x),ac(x))\ D^{-1/2}(x)\ d\lambda^u(x)\ ,$$

$$\hat{f}(x,a) = \int f(x,\xi)\ \overline{(\xi,ac(x))}\ d\xi, \text{ and}$$

$$\mathcal{F}\phi(u,a) = \int \phi(u,\gamma)\ \overline{(\gamma,a)}\ d\gamma\ .$$

This shows that $\|L(f)\| \le \|\pi(f)\|$ for every $f \in L^1(\hat{A},C^*(G,\sigma))$ and every induced representation L. Since $\hat{A}$ is abelian, the reduced norm on $L^1(\hat{A},C^*(G,\sigma))$ coincides with the C^*-norm ([70], proposition 2.2). Hence $\|f\| \le \|\pi(f)\|$.

Q.E.D.

5.8. Corollary : Let G be an r-discrete amenable principal groupoid with Haar system, $\sigma \in Z^2(G,\mathbb{T})$, A a locally compact abelian group and $c \in Z^1(G,A)$. Then the asymptotic range $R_\infty(c)$ of c coincides with the Connes spectrum $\Gamma(\alpha)$ of the corresponding automorphism group α on $C^*(G,\sigma)$.

Proof : We identify the crossed product C^*-algebra $C^*(G,\sigma) \times_\alpha \hat{A}$ and $C^*(G(c),\sigma)$. The canonical action of A on the skew product G(c), $s(a)(x,b) = (x,ab)$, defines an action on A on $C^*(G(c),\sigma)$, $\beta_a(f)(x,b) = f(x,a^{-1}b)$. Thus $C^*(G(c),\sigma),A,\beta)$ is nothing but the dual system of $(G\ (G,\sigma),\hat{A},\alpha)$. The Connes spectrum $\Gamma(\alpha)$ can be characterized as ([60], 8.11.8) $\Gamma(\alpha) = \{a \in A : J \cap \beta_a(J) \neq \{0\}$ for every non-zero ideal J of $C^*(G(c),\sigma)\}$. Using the correspondence 4.6 between ideals of $C^*(G(c),\sigma)$ and invariant open subsets of the unit space of G(c), the amenability of G(c) and 1.4.10, one gets the conclusion.

Q.E.D.

5.9. Remark : We have restricted our attention to automorphism groups of $C^*(G,\sigma)$ which stem from a cocycle $c \in Z^1(G,A)$. Another kind of automorphism group which leaves $C^*(G^0)$ invariant is given by a continuous action of a group A by automorphisms of G leaving the Haar system invariant and a similar study can be done.

CHAPTER III

SOME EXAMPLES

We shall give here two kinds of examples of r-discrete groupoids with Haar system. Our first example results from the observation by Strătilă and Voiculescu ([69], ch. I, §1, page 3) that approximately finite-dimensional C^*-algebras (for short AFC^*-algebras) could be diagonalized. This fact had already been used in a particular case be Gårding and Wightman in [34] to construct infinitely many non-equivalent irreducible representations of the anticommutation relations. In the terminology of 2.4.13, this can be rephrased by saying that AF C^*-algebras have Cartan subalgebras. Thus, an AF C^*-algebra is the C^*-algebra of an r-discrete principal groupoid. The groupoids which arise in that fashion (we call them AF) are studied in the first section. They have also been considered, in a form where the emphasis was on the ample group rather than on the groupoid, by Krieger in [52]. Our second example is given by the C^*-algebras generated by isometries introduced and studied by Cuntz in [15]. We show that these C^*-algebras may be written as groupoid C^*-algebras. The corresponding groupoids, which are described in the second section and which we call O_n, are not principal. In both cases, the description of the C^*-algebra in terms of a groupoid is used to discuss the existence of KMS-states with respect to some automorphism groups.

1. Approximately Finite Groupoids.

The simplest examples of r-discrete principal groupoids are, on one hand, the locally compact spaces (corresponding to the equivalence relation $u \sim v$ iff $u = v$)

and, on the other, the transitive principal groupoids on a set of n elements, where $n = 1,2,\ldots\infty$ (corresponding to the equivalence relation $u \sim v$ for every u and v) with the discrete topology. By means of elementary operations, we may combine them to obtain other examples.

The product of two groupoids is defined in the obvious fashion. If the groupoids are topological, then the product is given the product topology and if each of the groupoids is endowed with a Haar system, the product is given the product Haar system ; explicitly if $\{\lambda_i^{u_i}\}$ is a Haar system for G_i, $i = 1,2$, then $\{\lambda^{(u_1,u_2)} = \lambda_1^{u_1} \times \lambda_2^{u_2}\}$ is a Haar system for $G_1 \times G_2$.

Another operation makes sense in the category of groupoids ; this is the disjoint union. Let G_i be a groupoid, with $i = 1,2$; then define $G = G_1 \oplus G_2$ as the set-theoretical disjoint union of G_1 and G_2 with the groupoid structure given by the rules "x and y are composable in G iff they belong to the same G_i and are composable in G_i and their product in G is equal to their product in G_i" and "if x belongs to G_i, its inverse in G is equal to its inverse in G_i". If the groupoids G_i are topological, then their disjoint union is given the disjoint union topology and if $\{\lambda_i^{u_i}\}$ is a Haar system for G_i, $i = 1,2$, then $\{\lambda^u\}$, where $\lambda^u = \lambda_i^u$ if $u \in G_i^0$, is a Haar system for G. One can define in a similar fashion the disjoint union of a sequence of groupoids.

A last operation which we need here is the inductive limit. We give here a restricted definition, sufficient for our purposes. Suppose that the groupoid G is the union of an increasing sequence of subgroupoids G_n, which all have the same unit space as G ; then we say that G is the inductive limit of the sequence (G_n). If G is topological, we require that G_n be an open subgroupoid of G. If $\{\lambda^u\}$ is a Haar system for G, we consider the Haar system $\{\lambda_n^u\}$ on G_n such that λ_n^u is the restriction of λ^u to $r_n^{-1}(u)$. Conversely, suppose that the G_n's are topological groupoids such that G_n is open in G_{n+1} and its topology is the topology induced from G_{n+1}. Then, the inductive limit topology, where a set V is open iff $V \cap G_n$ is open in G_n for every n, makes G into a topological groupoid. If the G_n's are locally compact, then so is G. Finally, if each G_n has a Haar system $\{\lambda_n^u\}$ and if these measures are compatible, in the sense that λ_n^u is the restriction of λ_{n+1}^u to $r_n^{-1}(u)$, then there exists a unique Haar system $\{\lambda^u\}$ such that λ_n^u is the restriction of λ^u to $r_n^{-1}(u)$.

Let us note that these operations preserve amenability (definition 2.3.6). Let us show, for example, that the inductive limit G of a sequence (G_n) of amenable groupoids is amenable. Let K be a compact subset of G and ε a positive number. Since the G_n's are open, K is contained in some G_n. Since G_n is amenable, there exists $f \in C_c(G_n)$ such that $|f^* * f(x) - 1| \leq \varepsilon$ for $x \in K$ (and $\int |f(x)|^2 d\lambda_n^u$ bounded by 2). Then $f \in C_c(g)$ and satisfies the same condition in G.

1.1. Definition : Let G be an r-discrete groupoid. We say that G is an elementary groupoid of type n ($n = 1,2,\ldots,\infty$) if it is isomorphic to the product of a second countable locally compact space and of a transitive principal groupoid on a set of n elements.

We say that G is an elementary groupoid if it is the disjoint union of a sequence of elementary groupoids of G_i of type n_i.

We say that G is an approximately elementary (AE) groupoid if it is the inductive limit of a sequence of elementary groupoids.

We say that G is an approximately finite (AF) groupoid if it is approximately elementary and its unit space is totally disconnected.

1.2. Remarks : All these groupoids are principal and amenable since these properties are preserved under product, disjoint union and inductive limit. They have the counting measures as Haar system.

The orbits of an elementary groupoid of type n have the same cardinality n. However there exist r-discrete principal groupoids, all orbits of which have the same cardinality n, which are not elementary of type n. An example is given by the equivalence relation on the circle which identifies two points lying on the same diameter. The unit space of this groupoid is connected, while the unit space of an elementary groupoid of type 2 has at least two components.

The terminology of elementary groupoid does not agree with the definition (4.1.1) in [19]) of an elementary C^*-algebra. Only transitive principal groupoids give elementary C^*-algebras.

1.3. Proposition :

(i) Let G be an elementary groupoid. Then, for every G-module bundle A (not necessarily abelian), every cocycle $c \in Z^1(G,A)$ is inner, (that is, is a coboundary).

(ii) Let G be an approximately elementary groupoid. Then, for every G-module bundle A (not necessarily abelian), every cocycle $c \in Z^1(G,A)$ is approximately inner in the sense that it can be approximated by coboundaries uniformly on the compact subsets of G.

(iii) Let G be an approximately elementary groupoid. Then, for every abelian G-module bundle A and every $n \geq 2$, $H^n(G,A) = 0$.

Proof :

(i) We will show that an elementary groupoid is (continuously) similar to a locally compact space. Since a locally compact space (as a groupoid) has trivial cohomology, this will prove the assertion. It suffices to consider the case of an elementary groupoid of type n, of the form $G = X \times I_n$, where X is a locally compact space and I_n the transitive groupoid on $\{1,\ldots,n\}$. Then, a similarity between G and X is given by

$$\phi : X \times I_n \to X \quad (x,(i,j)) \mapsto x \qquad \text{and} \qquad \psi : X \to X \times I_n \quad x \mapsto (x,(1,1))$$

because $\phi\circ\psi = id_X$ and $\psi\circ\phi(x,(i,j)) = \theta(x,i)id_G(x,(i,j))\theta(x,j)^{-1}$ where θ is the map

$$X \times \{1,\ldots,n\} \to X \times I_n, \quad (x,i) \mapsto (x,(1,i)).$$

(ii) Let G be the inductive limit of a sequence of elementary groupoids G_n and let $c \in Z^1(G,A)$. By (i), the restriction $c_{|G_n}$ of c to G_n is a coboundary on G_n, hence may be extended to a coboundary c_n on G. Since every compact subset of G is contained in some G_n, (c_n) converges to c uniformly on the compact subsets of G.

(iii) Write G as increasing union of a sequence of elementary groupoids G_i. Let $\sigma \in Z^n(G,A)$, with $n \geq 2$. Its restriction to G_i, σ_i, belongs to $Z^n(G_i,A)$. However $Z^n(G_i,A) = (0)$ for $n \geq 2$, since $Z^m(G_i,A) = B^m(G_i,A)$ for $m \geq 1$. Thus $\sigma = 0$.

Q.E.D.

An essential feature of an approximately elementary groupoid G is that it has (c,β) KMS measures for every $c \in Z^1(G,\mathbb{R})$ and every $\beta \in [-\infty,+\infty]$, provided that its unit space is compact.

1.4. Lemma : Let G be a locally compact groupoid with Haar system and let c be a coboundary in $B^1(G,\mathbb{R})$.

(i) If G^0 is compact, then (c,∞) KMS probability measures exist.

(ii) If there is a (c,β) KMS measure for some $\beta \in \mathbb{R}$, then there are (c,β') KMS measures for every $\beta' \in \mathbb{R}$.

(iii) If G^0 is compact and if there is a (c,β) KMS probability measure for some $\beta \in \mathbb{R}$, then there are (c,β') KMS probability measures for every $\beta' \in [-\infty,+\infty]$.

Proof : Let us write $c(x) = h \circ r(x) - h \circ d(x)$ where h is a continuous function on G^0.

(i) The set Minh of the points of G^0 where h reaches its minimum is non-empty and contained in Minc. The point-mass at such a point is a (c,∞) KMS probability measure.

(ii) If μ is a (c,β) KMS measure, then for every $\beta' \in \mathbb{R}$, the measure μ' given by

$$d\mu'(u) = \exp[-(\beta'-\beta)\, h(u)] \quad d\mu(u)$$

is a (c,β') KMS measure. For, if $\nu' = \int \lambda^u d\mu'(u)$,

$$\frac{d\nu'}{d\nu'^{-1}}(x) = \exp[-(\beta'-\beta)h\circ r(x)]\ \frac{d\nu}{d\nu^{-1}}(x) \exp\,[(\beta-\beta')h\circ d(x)]$$

$$= \exp[-\beta c(x)].$$

(iii) If μ, as above, is finite and if G^0 is compact, μ' is also finite.

Q.E.D.

..5. Proposition : Let G be an approximately elementary groupoid with compact unit pace. Then it admits (c,β) KMS probability measures for every $c \in Z^1(G,\mathbb{R})$ and every $\varepsilon\ [-\infty,+\infty]$.

roof : Since elementary groupoids with compact unit space have finite invariant easures, they have (c,β) KMS probability measures for every c and every β. Fix $\in\mathbb{R}$ and $c \in Z^1(G,\mathbb{R})$. Write G as the inductive limit of a sequence (G_n) of elementary roupoids and let c_n be the restriction of c to G_n. For each n, there exists a probability measure μ_n whose modular function with respect to G_n is $e^{-\beta c_n}$. Let μ be a limit point of the (μ_n)'s for the weak $*$-topology of the dual of the space of continuous functions on G^0. If $\mu_n \to \mu$, then $\nu_n \to \nu$ and $\nu_n^{-1} \to \nu^{-1}$ for the weak $*$-topology of the dual of $C_c(G)$. Therefore, for every $f \in C_c(G)$,

$$\int f d\nu^{-1} = \lim \int f d\nu_n^{-1} = \lim \int f e^{\beta c_n} d\nu_n = \int f e^{\beta c} d\nu.$$

This shows that the modular function of μ exists and is $e^{-\beta c}$.

The statement about infinite β results from 1.3.17.

Q.E.D.

1.6. Example : The Ising model.

The points of $Z = \mathbb{Z}^{\nu}$ are the sites of a crystal lattice of dimension ν, where ν is an integer.Each site has a spin up (-1) or down (-1). A configuration of the lattice is given by a function u of Z into $\{-1,+1\}$. The space of configuration $\{-1,+1\}^Z$ is given the product topology ; it will be the unit space G^0 of the groupoid. Two configurations are equivalent iff they differ at most finitely many sites. The corresponding principal groupoid is noted G. We choose an increasing sequence (Z_n) of finite subsets of the lattice such that $Z = \cup Z_n$ and define the subgroupoid G_n by the equivalence relation : "two configurations are equivalent if they agree outside Z_n". Then G_n is an elementary groupoid of the form $\{-1,+1\}^{Z\setminus Z_n} \times I_{[Z_n]}$ and $G = \cup G_n$. We give to G the inductive limit topology. Thus G is an AF groupoid.

The dynamics of the system are described by the following energy cocycle $c \in Z^1(G, \mathbb{R})$ given by the expression

$$c(u,v) = \sum_{i,j} J(i,j) \{(1 - u_i u_j) - (1 - v_i v_j)\},$$

where J depends on the nature of the interaction. The sum is in fact finite since there are finitely many non zero terms.

From 1.5, the system has KMS states for every β. The ground states are the measures which live on $\{u \in G^0 : u_i u_j = 1 \text{ whenever } J(i,j) \neq 0\}$. In particular, the configurations ($u_i = +1$ for every i) and ($u_i = -1$ for every i) are physical ground states.

Some results, depending on β and on J, are known above the existence of distinct KMS states at a given β.The parameter β is interpreted as the inverse temperature and KMS states are equilibrium states. Coexistence of distinct KMS states means the existence of several "phases". If the lattice were finite, G would be finite, c inner and there would be one and only one KMS state for every β. The interested reader will find a review of these results as well as a bibliography in the A.M.S. article by J. Fröhlich [33].

We turn now to the properties of the skew-product $G(c)$ where G is approximately elementary (or finite).

1.7. Proposition : Let G be a locally compact groupoid, A a locally compact group and c a cocycle in $Z^1(G,A)$.

(i) If G is approximately elementary, then the skew product $G(c)$ of G by c is approximately elementary.

(ii) If G is approximately finite and A is totally disconnected, then $G(c)$ is approximately finite.

Proof :

(i) If c is a coboundary, $c(x) = b\circ r(x)\,(b\circ d(x))^{-1}$. Then, the map from $G \times A$ to $G(c)$ sending (x,a) to $(x, a(b\circ r(x))^{-1})$ is an isomorphism of groupoids, when $G \times A$ is given the product structure and where A is viewed as a locally compact space. Therefore, if G is elementary, $G(c)$ is also elementary for every $c \in Z^1(G,A)$.

Suppose now that $G = \cup G_n$ with G_n elementary. Let $c \in Z^1\,(G,A)$ and let c_n be its restriction to G_n. Then $G(c) = \cup G_n(c_n)$ and $G_n(c_n)$ is elementary. Thus, by definition, $G(c)$ is approximately elementary.

(ii) From the first part, we know that $G(c)$ is approximately elementary. Moreover its unit space $G^0 \times A$ is totally disconnected. Hence it is approximately finite.

Q.E.D.

Remark : This last proposition gives a partial answer to a question Bratteli asks in [9] (problem 2, page 35). If $(\mathcal{A},G,\alpha)$ is a C^*-dynamical system with $\mathcal{A}$ AF and G compact, is the crossed product algebra $\mathcal{A}\times_\alpha G$ necessarily AF ? This is so if G is abelian and the action is given by a cocycle as in 2.5.1.

The crossed-products of UHF algebras by product-type actions studied by Bratteli in [9] are aptly described in terms of groupoids. Let (X_i) be a sequence of finite discrete spaces and let $X = \prod X_i$ be their product, with the product topology. The equivalence relation $\sim$ on X, where $u \sim v$ iff $u_i = v_i$ for all but a finite number of indices, defines a principal groupoid G. If the sequence is indexed by $\mathbb{N}$ we may define the groupoid $G^i = \{(u,v) \in G : u_j = v_j \text{ for } j \geq i\}$, which is elementary. As in example 1.6, $G = \bigcup G^i$ is made into a topological groupoid which is AF. Since every point of $G^0 = X$ has a dense orbit, G is minimal. A topological groupoid isomorphic to such a groupoid G will be called a Glimm groupoid, because, as we shall see, its C^*-algebra is a UHF, or Glimm, algebra.

Let A be an abelian locally compact group. A cocycle $c \in Z^1(G,A)$ will be said of product type if it is of the form

$$c(u,v) = \sum c_i(u_i,v_i) \text{ where } c_i \in Z^1(G_i,A)$$

where G_i is the transitive groupoid on the set X_i. We may write

$$c_i(u_i,v_i) = b_i(u_i) - b_i(v_i)$$

with b_i function from X_i into A. We let $C_i = c_i(G_i) = B_i - B_i$ where $B_i = b_i(X_i)$. We may assume that $0 \in B_i$. Let us note that, by the definition of the topology of G as inductive limit topology, a cocycle of product type is continuous.

<u>1.8. Proposition :</u> Let G be a Glimm groupoid, A an abelian locally compact group and let c be a cocycle in $Z^1(G,A)$ of product type as above.

(i) The asymptotic range of c is $R_\infty(c) = \bigcap_{j\in\mathbb{N}} \overline{(\sum_{i\geq j} C_i)}$.

(ii) Its T-set is $T(c) = \{\xi \in \hat{A} : \forall \varepsilon > 0, \ j : \ |\xi(\sum_{i\geq j} B_i) - 1| \leq \varepsilon\}$.

(iii) The cocycle c is a coboundary iff for every neighborhood V of 0 in A, there exists j such that $\sum_{i\geq j} B_i$ is contained in V.

(iv) The asymptotic range of c at u is $R_\infty^u(c) = \bigcap_{j\in\mathbb{N}} \overline{(\sum_{i\geq j} B_i - b_i)}$ where $b_i = b_i(u_i)$.

<u>Proof :</u> The assertions (i), (ii), and (iv) result from the definition 1.4.3. The assertion (iii) results from proposition 1.4.8. We have to check that the hypotheses of this proposition are satisfied. The unit space of G is compact and G admits a cover of continuous G-sets, namely, the sets

$$S = \{((a,u^i),(\sigma(a),u^i)) \in G : a \in \prod_1^{i-1} X_j,\ u^i \in \prod_{j\geq i} X_j\}$$

where i is an integer and σ a bijection of $\prod_1^{i-1} X_j$ onto itself.

Q.E.D.

Bratteli points out in [9] that the simplicity of the crossed-product algebras he considers depends heavily on the structure of the group A. This is summarized in the following proposition.

1.9. Proposition : Let G be a Glimm groupoid, A an abelian locally compact group and let c be a cocycle in $Z^1(G,A)$ of product type.

(i) If A is compact and $R_\infty(c) = A$, then G(c) is minimal

(ii) If A can be ordered, then G(c) is not minimal.

Proof : The assertion (i) results directly from 1.4.16 (ii).

To prove the second assertion, we use the notation given above. We may choose b_i so that $B_i = b_i(X_i)$ is contained in the positive cone P of A and b_i is non-decreasing when $X_i = \{0,1,\dots,n_i\}$ has its usual order. Let 0 and 1 denote respectively the sequences $0 = (0,0,\dots,)$ and $1 = (n_1,n_2,\dots)$ in $X = \prod X_i$. Then the asymptotic range at 0 of c, $R_\infty^0(c) = \bigcap_{j\in\mathbb{N}} \overline{(\sum_{i\geq j} B_i)}$, is contained in the positive cone of A while the asymptotic range at 1 of c, $R_\infty^1(c) = \bigcap_{j\in\mathbb{N}} \overline{(\sum_{i\geq j} - B_i)}$ is contained in the negative cone of A, -P. By 1.4.14 (i), for every $a \in A$, the points (0,a) and (1,a) do not have a dense orbit. Therefore, G(c) is not minimal.

Q.E.D.

1.10. Example : The gauge automorphism group of the CAR algebra.

Let us first define the CAR groupoid (CAR stands for canonical anticommutation relations). It is a Glimm groupoid isomorphic to the groupoid of the Ising model. The positions of a system of fermions are labeled by a countable set of indices, say $\mathbb{N}$. The unit space of the groupoid is $X = \prod_{i\in\mathbb{N}} X_i$ where $X_i=\{0,1\}$. A configuration $u=(u_i)$ in X tells if there is a fermion at the place i. As before, G is the principal groupoid given by the equivalence relation $\sim$, where two configurations are equivalent iff they differ at at most a finite number of places. We shall see that its C^*-algebra is the C^*-algebra of the canonical anticommutation relations (see [8] or [29] page 269).

The gauge automorphism group is defined by the product cocycle $c \in Z^1(G,\mathbb{Z})$, called the "number" cocycle, given by

$$c(u,v) = \sum_{i\in} u_i - v_i \ .$$

The number cocycle counts the number of particles by which the configurations u and v differ. With above notations, $B_i = \{0,1\}$ and $R_\infty(c) = \mathbb{Z}$.

Let us define next the GICAR groupoid (GI stands for gauge invariant). It is the subgroupoid $c^{-1}(0)$. In other words, it corresponds to the equivalence relation $\approx$, where two configurations u and v are equivalent iff they differ at at most a finite number of places and have the same number of particles (in the sense that $c(u,v) = 0$). Its C^*-algebra is the subalgebra of fixed points of the gauge automorphism group ; it is called the GICAR algebra. It results from 1.4.17 that the GICAR groupoid is irreducible. More information about it will be given after we introduce the dimension group of an AF-groupoid.

Finally, let us consider the skew-product groupoid $G(c)$. By 1.7 it is an AF-groupoid. It is irreducible (by 1.4.13) but not minimal (by 1.9).

The remainder of this section is devoted exclusively to topological groupoids which admit a base of open sets consisting of compact open G-sets. After a few definitions, we shall study the example of AF-groupoids.

Let G be a topological groupoid which admits a base of compact open G-sets. Its ample semi-group $\mathcal{G}$ has been defined (1.2.10) as the inverse semi-group of its compact open G-sets. The idempotent elements of $\mathcal{G}$ are compact open subsets of the unit space G^0 of G. They form a generalized Boolean algebra $\mathcal{G}^0$ (that is, a Boolean algebra without the assumption that a greatest element exists). We define the following equivalence relation on $\mathcal{G}^0$. We shall declare e and f equivalent, and write $e \sim f$, iff there exists $s \in \mathcal{G}$ such that $e = r(s)$ and $f = d(s)$, where $r(s) = ss^{-1}$ and $d(s) = s^{-1}s$.

Using terminology common to the theory of von Neumann algebras, one can make the following definition.

1.11. Definition : Let G be a topological groupoid which has a base of compact open G-sets and let $\mathcal{G}$ be its ample semi-group. We say that an idempotent element e of $\mathcal{G}$ is finite if for any idempotent element f, the relation $e \sim f \leq e$ implies $f = e$. We say that G is of finite type if every idempotent element of $\mathcal{G}$ is finite and of infinite type otherwise.

We may define on $\mathcal{G}^0$ the relation $e \prec f$ iff there exists e_1 and f_1 such that $e \sim e_1 \leq f$. We may also define a partial addition in $\mathcal{G}^0$, where two idempotent elements e and f can be added iff they are disjoint and e + f is the union of e and f. We denote by D(G) the of equivalence classes $\mathcal{G}^0/\sim$ and by D the quotient map of $\mathcal{G}^0$ onto D(G). We provide D(G) with the relation $D(e) \leq D(f)$ iff $e \prec f$ and with a partial addition, where two classes D(e) and D(f) can be added iff they contain disjoint elements e_1 and f_1 and then $D(e) + D(f) = D(e_1 + f_1)$. If G is of finite type, the relation $\leq$ is an order relation.

1.12. Definition : Let G be a topological groupoid which admits a base of compact open G-sets and let $\mathcal{G}$ be its ample semi-group. Assume that G is of finite type. Then, its dimension range is the set $D(G) = \mathcal{G}^0/\sim$ with the order structure and the partial additive structure defined as above.

It can be shown (cf. [25] and [27] in the AF case) that the dimension range D(G) of G can be embedded in a unique fashion as a generating upward directed hereditary subset of a directed ordered abelian group, called the dimension group of G and denoted by $K_0(G)$.

The property of being of finite type is preserved under finite products, disjoint unions and inductive limits. It can be shown that

$$K_0(G_1 \times G_2) = K_0(G_1) \otimes K_0(G_2),$$

with positive cone generated by

$$K_0^+(G_1) \otimes K_0^+(G_2) \; ;$$

$$D(G_1 \times G_2) = \{\sum_1^n m_i \times n_i \; ; \; n \in \quad , m_i \in D(G_1),\ n_i \in D(G_2), \sum_1^n m_i \in D(G_1), \sum_1^n n_i \in D(G_2)\}$$

$$K_0(\oplus G_n) = \sum K_0(G_n) ,$$

where $\oplus G_n$ is the disjoint union of a sequence (G_n) and $\sum K_0(G_n)$ is the direct sum of the ordered abelian groups $K_0(G_n)$;

$$D(\oplus G_n) = \{\sum_1^k m_i : k \in \mathbb{N}, m_i \in D(G_i)\},$$

$$K_0(\varinjlim G_n) = \varinjlim K_0(G_n) ,$$

where $\varinjlim G_n$ is the inductive limit of an increasing sequence (G_n) and $\varinjlim K_0(G_n)$ is the inductive limit of the ordered abelian groups $K_0(G_n)$; and

$$D(\varinjlim G_n) = \cup D(G_n) .$$

An ordered abelian group will be called an <u>Elliott group</u> (cf. [25], [27] and [28]) if it is the inductive limit of a sequence of ordered groups, each isomorphic to the direct sum of finitely many copies of $\mathbb{Z}$ with its usual order. The property of being an Elliott group is preserved under finite tensor products, countable direct sums and countable inductive limits.

For example, the dimension group of a second countable totally disconnected locally compact space X is an Elliott group. Indeed, the dimension range of X is the (generalized) Boolean algebra of its compact open subsets, which may be written as an increasing union of a sequence of finite Boolean algebras $\mathcal{B}_n$ and its dimension group is the group of continuous functions with compact support of X into $\mathbb{Z}$, $C_c(X,\mathbb{Z}) = \{f \in C_c(X,\mathbb{Z}) : \text{supp} f \in \mathcal{B}_n\}$, with its usual order. The dimension range of the transitive groupoid on a set of n elements, where $n = 1,2,\ldots,\infty$, is $\{0,1,\ldots,n\}$ and its dimension group, which is $\mathbb{Z}$, is an Elliott group. Therefore, the dimension group of an AF groupoid is an Elliott group.

The importance of the dimension ranges and of the dimension groups in the study of AF groupoids is given by the following proposition. The first assertion is essentially proposition 3.3 of [52] and the second assertion is theorem 3.5 of [52]. Let us note that this theorem has a long history in the context of C^*-algebras ([35], [18], [8] and [27]).

<u>1.13. Proposition</u> (W.Krieger) :

(i) The dimension group of an AF-groupoid is an Elliot group and every Elliott group occurs as the dimension group of an AF groupoid.

(ii) Two AF-groupoids are isomorphic iff their dimension ranges are isomorphic.

The AF-groupoids considered in [52] have a compact unit space, but, as pointed out there, this assumption can be removed. Given an AF-groupoid G with compact unit space, the subgroup of its ample semi-group consisting of those G-maps which are everywhere defined is an ample CLF group in the sense of [52]. Conversely, given an ample CLF group acting on the space X, the groupoid of the corresponding equivalence relation on X is AF.

Let us describe the dimension range of the AF-groupoids that we have met in this section. The dimension group of the Glimm groupoid of the Ising model (and of the canonical anticommutation relations) is the group $Q(2^{\infty})$ of rational numbers whose denominator is a power of 2, with the order inherited from $\mathbb{Q}$. Its dimension range is the segment $[0,1]$. With the notations of 1.6, the dimension of a cylinder set $C(Z_n)$ obtained by fixing the spins inside a finite subset Z_n of the lattice is $D(C(Z_n)) = 2^{-|Z_n|}$. There exists a unique probability measure μ on $\{-1,+1\}^Z$ which extends D. It is the unique ergodic invariant probability measure of the groupoid.

We give in the appendix a computation of the dimension group of the GICAR groupoid. It is the group $\mathbb{Z}[t]$ of polynomials in one variable with integer coefficients, where the order is given by $f > 0$ iff $f(t) > 0$ for every $t \in]0,1[$. This is an example of a Riesz group (cf. [25]). There are uncountably many invariant ergodic probability measures, indexed by $t \in]0,1[$ and obtained by composing the dimension map with the point evaluation at t. The measure corresponding to $t = \frac{1}{2}$ is the unique invariant probability measure for the CAR groupoid.

The dimension group of the skew-product of the CAR groupoid and the number cocycle can be computed in the same fashion as the dimension group of the GICAR roupoid. It is the group $\mathbb{Z}(t)$ of rational functions with integer coefficients and hose only possible poles are at 0 and 1, where the order is given by $f > 0$ iff $(t) > 0$ for every $t \in]0,1[$.

Let us look at the relationship between AF-groupoids and AF C^*-algebras. It is ue to Krieger ([52], theorem 4.1) and relies essentially on a result of Strătilă and Voiculescu ([69], section 1 of chapter 1). We give a self-contained proof which is essentially the same as theirs. Let us recall that an AF C^*-algebra is the inductive limit of a sequence of finite-dimensional C^*-algebras. Basic references for AF C^*-algebras are [35], [18] and [8].

The crux of the proof is the following lemma about finite-dimensional C^*-algebras.

1.14. Lemma : Let A be a finite-dimensional $*$-algebra and A_1 a sub $*$-algebra. Then, for any Cartan subalgebra B_1 of A_1, there exists a Cartan subalgebra B of A which contains B_1 and whose normalizer $\mathcal{N}(B)$, that is, the inverse semi-group of partial isometries a of A such that $d(a), r(a) \in B$ and $a(Bd(a))a^* = Br(a)$, contains the normalizer $\mathcal{N}_1(B_1)$ of B_1 in A_1.

Proof : Since A_1 is a sum of simple $*$-algebras, we may assume that A_1 itself is simple. The normalizer $\mathcal{N}_1(B_1)$ of B_1 in A_1 contains matrix units (e_{ij}) $i,j = 1,\ldots,m$ which span A_1. The projection e_{11} of B_1 decomposes in A into minimal projections : $e_{11} = f_1 + \ldots + f_n$. The family $(e_{i1} f_j e_{1i})$ $i = 1,\ldots,m$ and $j = 1,\ldots,n$ consists of orthogonal projections and is contained in a Cartan subalgebra B of A. The algebra B_1, which is spanned by the projections (e_{ii}) $i = 1,\ldots,m$, is a subalgebra of B. The matrix units (e_{ij}) normalize B. Therefore $\mathcal{N}_1(B_1)$ is contained in $\mathcal{N}(B)$

Q.E.D.

1.15. Proposition : Let A be a C^*-algebra. The following properties are equivalent.

(i) The C^*-algebra A is AF.

(ii) The C^*-algebra A is the C^*-algebra of an AF-groupoid G. Moreover, under these conditions, the AF-groupoid G is unique up to isomorphism and its dimension range is the dimension range of A (cf. [27]).

Proof : Suppose that A is an AF C^*-algebra and choose an increasing sequence of finite-dimensional C^*-algebras A_n which defines A. Construct by induction a sequence of Cartan subalgebras B_n of A_n such that B_{n+1} contains B_n and its normalizer $\mathcal{N}_{n+1}$ in A_{n+1} contains the normalizer $\mathcal{N}_n$ of B_n in A_n. Let B be the closure of the union of the B_n's. Since $\mathcal{N}_n$ normalizes B_m for $m \geq n$, it normalizes B, hence the ample inverse semi-group $\mathcal{G}_n$ of B_n acts on B. We realize B as $C^*(X)$, where X is a totally disconnected locally compact space and we let $\mathcal{G} = \cup \mathcal{G}_n$, viewed as an inverse semi-group of partial homeomorphisms of X. The corresponding equivalence relation on X yields a principal groupoid G which is AF because it is of the form $G = \cup G_n$, where G_n is the principal groupoid of the equivalence relation corresponding to $\mathcal{G}_n$. It is almost obvious that G_n is an elementary groupoid. For, $\mathcal{G}_n$ partitions the atoms of the Boolean algebra $\mathcal{B}_n$ of projections of B_n into equivalence classes. Let $\{Y_1,\ldots,Y_m\}$ be one of these classes and let $Y = Y_1 \cup \ldots \cup Y_m$. Then the reduction of G_n to Y is isomorphic to $Y_1 \times I_m$, where I_m is the transitive groupoid on m elements. The lemma allows the construction of consistent systems of matrix units in each algebra A_n. In other words, there exists a section k for the canonical map of $\mathcal{N} = \cup \mathcal{N}_n$ onto $\mathcal{G}$. Let $C^*(\mathcal{G}_n)$ be the (finite-dimensional) sub C^*-algebra of $C^*(G)$ generated by $\{x_s : s \in \mathcal{G}_n\}$. There exists an isomorphism ϕ_n of $C^*(\mathcal{G}_n)$ into A_n such that $\phi_n(x_S) = k(S)$ for $S \in \mathcal{G}_n$. Since the restriction of ϕ_{n+1} to $C^*(\mathcal{G}_n)$ is ϕ_n, there exists an isomorphism ϕ of $\bigcup_n C^*(\mathcal{G}_n)$ onto $\bigcup_n A_n$ whose restriction to $C^*(\mathcal{G}_n)$ is ϕ_n. It is isometric with respect to the C^*-norms of $C^*(G)$ and of A, because finite-dimensional $*$-algebras have a unique C^*-norm. Therefore, it extends to an isomorphism of $C^*(G)$ onto A.

The above argument also shows that the C^*-algebra of an AF-groupoid is AF.

Let us keep the same notations as above. The dimension range $D(\mathcal{G}_n) = \mathcal{B}_n/\mathcal{G}_n$ is also the dimension range $D(A_n)$ of the $*$-algebra A_n. The dimension range of G, which is the inductive limit of the dimension ranges $D(\mathcal{G}_n)$, is equal to the dimension range of the locally finite $*$-algebra $\cup A_n$. It is known (e.g. [27], remark 4.4, page 34) that this is also the dimension range of A. Therefore, the uniqueness of the AF-groupoid G results from 1.13 (ii).

Q.E.D.

.16. Corollary : Suppose that a C^*-algebra A has two Cartan subalgebras B_1 and B_2 which are both AF and which have countable locally finite ample semi-groups, then B_1 and B_2 are conjugate by an automorphism of A.

Proof : The groupoids G_1 and G_2 obtained by 2.4.15 are AF. (Therefore, the 2-cocycles σ_1 and σ_2 are equal to 1). By the previous proposition, A is AF and G_1 and G_2 have the same dimension range. Therefore, they are isomorphic and an isomorphism of G_1

onto G_2 implements an automorphism of A carrying B_1 onto B_2.

Q.E.D.

This is the only result we have about the existence and the uniqueness of Cartan subalgebras. It is not known, even in the case of an AF C^*-algebra, whether a C^*-algebra may have non-conjugate Cartan subalgebras.

The following example shows that the definition we give of a Cartan subalgebra cannot be weakened if we expect uniqueness.

Let K be the algebraic closure of a finite field, with the discrete topology. The multiplicative group of K is denoted K^*, its additive group is denoted K^+ and the dual group of K^+ is denoted $\hat{K}^+$. Since K is an increasing sequence of finite fields K_n, K^+ is the inductive limit of finite groups K_n^+ and $\hat{K}^+$ is the projective limit of finite groups $\hat{K}_n^+$. As a topological space $\hat{K}^+$ is homeomorphic to the Cantor space.

The "ax + b" group over K is the semi-direct product $G = K^+ \times_\alpha K^*$, where K^* acts on K^+ by multiplication. It is equipped with the product topology. We view K^+ as a normal abelian subgroup of G. Since G has the discrete topology, the C^*-algebra $B = C^*(K^+)$ is a subalgebra of $A = C^*(G)$.

<u>1.17. Proposition :</u> Let A and B be as above.

(i) The C^*-algebra A is AF.

(ii) The subalgebra B is maximal abelian, regular, is the image of a unique (faithful) conditional expectation but its ample semi-group does not act relatively freely on the spectrum $\hat{K}^+$ of B, hence it fails to be a Cartan subalgebra.

<u>Proof :</u> (cf. Dixmier [17]).

(i) As above, we write K as union of an increasing sequence of finite fields K_n. The "ax + b" group over K_n, G_n, is a subgroup of G and G is the union of the G_n's. As in 1.15, we see that $C^*(G)$ is the inductive limit of the $C^*(G_n)$'s, which are finite-dimensional.

(ii) As an increasing union of finite groups, G is amenable. We apply 2.4.2. to view the elements of $C^*(G)$ as functions on G vanishing at infinity. The elements of

$C^*(K^+)$ are those functions which vanish outside K^+. To show that B is maximal abelian, we pick an element f of its commutant in A. It satisfies $\varepsilon_{b_1} * f * \varepsilon_{-b_1} = f$ for every $b_1 \in K^+$, where ε_{b_1} is the point mass at b_1. Explicitly, this gives $f(a,(1-a)b_1 + b) = f(a,b)$ for every $b_1 \in K^+$, $a \in K^*$, $b \in K^+$. Since f vanishes at infinity, this is only possible if $f(a,b) = 0$ when $a \neq 1$, that is, $f \in B$.

Since K^+ is a normal subgroup, the normalizer of B contains the elements ε_x, where $x \in G$. Therefore B is regular.

Let P be a conditional expectation onto B. From the relations

$(1,b_1)\,(a,b)\,(1,b_1)^{-1} = (a,(1-a)b_1 + b)$ and

$(1,b_1)\,(a,b) = (a, b + b_1)$

for every $a \in K^*$ and every $b, b_1 \in K^+$, we obtain that $P(\varepsilon_{(a,b)}) = P(\varepsilon_{(a,(1-a)b_1+b)}) = \varepsilon_{(1-a)b_1} * P(\varepsilon_{(a,b)})$. Thus, if $a \neq 1$, $P(\varepsilon_{(a,b)})$ is invariant under translation. Since it vanishes at infinity, it must be zero. This shows that the restriction of P to $C_c(G)$ is the restriction map of $C_c(G)$ onto $C_c(K^+)$. On the other hand, it results from 2.2.9 that this restriction map is positive and bounded. Hence it extends uniquely to a conditional expectation of $C^*(G)$ onto $C^*(K^+)$, which is still given by restricting a function to K^+. It is clearly faithful.

To show that the ample semi-group of B does not act relatively freely on $\hat{K}^+$, we note that the element $\varepsilon_{(a,b)}$ of the normalizer of B induces the homeomorphism s_a of its spectrum $\hat{K}^+$, where $s_a(\chi) = a\chi$ and $a\chi(b) = \chi(ab)$ for $\chi \in \hat{K}^+$. If $a \neq 1$, the set of fixed points of s_a is reduced to the identity character 1, hence is not open in $\hat{K}^+$.

Q.E.D.

We have not been able to determine whether the exact sequence

$$\mathcal{B} \to \mathcal{U}(B) \to \mathcal{N}(B) \overset{S}{\to} \mathcal{G}(B) \to \mathcal{B}$$

splits or not.

.18. Remark : The C^*-algebra A is the C^*-algebra of the transformation group $(\hat{K}^+, K^*)$ where the action of K^* on $\hat{K}^+$ is described above. Since $Y = \hat{K}^+ \setminus \{1\}$ is an invariant open subset of $\hat{K}^+$, A is an extension of $C^*(Y \times_\alpha K^*)$ by $C^*(K^*)$ (2.4.4). One can show that the dimension range of the groupoid $Y \times K^*$ is the segment $[0,p[$

of the dimension group $Q(p^\infty)$ of rational numbers whose denominator is a power of p, where p is the characteristic of the field K. Therefore, $C^*(Y \times K^*)$ is a matroid algebra without unit of type $M_{p^\infty,p}$ (notation of [18]). On the other hand, the C^*-algebra $C^*(K^*)$ is the C^*-algebra of the Cantor space. It results from [25], section 5.1, that the dimension group of A is an extension of $Q(p^\infty)$ by the dimension group of the Cantor space.

Let us mention here, without giving the details, that such extensions are characterized, up to equivalence, by measures on the Cantor space. Explicitly, one finds that $K_0(A) = Q(p^\infty) \times C(\hat{K}^*, \mathbb{Z})$ and that an element (q,f) is positive if and only if $q + \phi(f)$ is positive where ϕ is the measure on K^*, constructed as follows. Let (n_i) be a sequence of integers such that n_i divides n_{i+1} and $n_1 = 1$, let $q_i = p^{n_i}$ and let $f_i = \dfrac{q_i - 1}{q_{i-1}-1}$ for $i \geq 2$ and $f_1 = p$. Realize the space $\hat{K}^*$ as the product space $\prod_{i=1}^{\infty} \{1,2,\ldots,f_i\}$. The measure ϕ is concentrated on the points (a_i) with $a_i = 1$ for i large enough. If k is the last index i for which $a_i \neq 1$ (or, if $a_i = 1$ for every i, set $k = 1$), the measure of the point (a_i) is $\sum_{j=k}^{\infty} \dfrac{p}{q_{j-1}} \left[\dfrac{1}{q_j} - \dfrac{1}{q_{j+1}}\right]$.

The dimension range of A is the segment $[0,c]$, where c is the element $(p-1,1)$ of $Q(p^\infty) \times C(\hat{K}^*, \mathbb{Z})$.

Another method to check that the subalgebra B is not a Cartan subalgebra is to determine its dimension range and its dimension group relative to A. Its dimension group is an extension of $Q(p^\infty)$ by $\mathbb{Z}$.

2. The Groupoids O_n

The aim of this section is to exhibit the C^*-algebras generated by isometries introduced by J.Cuntz in [15] as the C^*-algebras of a groupoid. The groupoids we construct are not principal and we do not know if these algebras can be realized as the C^*-algebras of a principal groupoid. Nevertheless, this description of the Cuntz algebras reveals much of their structure. It also makes apparent the relationship between these algebras and some inverse semi-groups.

We start with a crossed product construction prompted by the representation of the Cuntz algebras as a crossed product (section 2 of [15]). We include the case $n = 1$, which will give the algebra of the bicyclic semi-group, studied by Barnes in [1].

For every $n = 1,2,\ldots,\infty$, we define the following AF-groupoids G_n. The groupoid G_1 is the compact space $\overline{\mathbb{Z}} = \mathbb{Z} \cup \{\infty\}$, one-point compactification of the space of integers $\mathbb{Z}$ with its discrete topology. We recall that it corresponds to the equivalence relation $u \sim v$ iff $u = v$ on $\overline{\mathbb{Z}}$.

For n larger than 1 but finite, the groupoid G_n corresponds to the equivalence relation $u \sim v$ iff $u_i = v_i$ for all but finitely many i's on the compact space $\{0,1,\ldots,n-1\}^{\mathbb{Z}}$ with the product topology. This is a Glimm groupoid. Its dimension group is the group $\mathbb{Q}(n^\infty)$ of rational numbers whose denominator is a power of n, with the order inherited from $\mathbb{Q}$ and its dimension range is the segment $[0,1]$.

The unit space of the groupoid G_∞ is the space $G_\infty^0 = \{u \in \overline{\mathbb{N}}^{\mathbb{Z}} : u_i = 0$ for i sufficiently small and $u_j = \infty$ for every $j \geq i$ if $u_i = \infty\}$, where $\overline{\mathbb{N}} = \mathbb{N} \cup \{\infty\}$. The cylinder sets $Z(\alpha)$, where $\alpha = (\ldots,0,j_k,\ldots,j_{k+\ell})$ with $k \in \mathbb{Z}$, $\ell \in \mathbb{N}$ and $j_{k+i} \in \mathbb{N}$, and their complements form a subbase of open sets for a topology on G_∞^0. This topology is locally compact and totally disconnected. The cylinder sets $Z(\alpha)$ are compact. We define, for $u \in G_\infty^0$, $k(u)$ as the smallest index i such that $u_i = \infty$, if it exists and as ∞ if $u_i < \infty$ for every i. The groupoid G_∞ corresponds to the equivalence relation on G_∞^0 : $u \sim v$ iff $k(u) = k(v)$ and $u_i = v_i$ for all but finitely many i's. One checks as in the example of the Glimm groupoids that it is an AF-groupoid. The closure of the orbit of a point u is $[\bar{u}] = \{v : k(v) \leq k(u)\}$. In particular, there are dense orbits. The invariant open sets form a decreasing sequence (U_i), $i \in \mathbb{Z}$, where $U_i = \{u : k(u) \geq i\}$. The dimension group of the groupoid G_∞ is the lexicographical direct sum $\sum_{i\in\mathbb{Z}} \mathbb{Z}$ (cf. 5.3 of [28]) and its dimension range is the whole positive cone. Further references to the AF C^*-algebras whose dimension group is totally ordered can be found in [28].

In each case, there exists a natural shift ϕ^0 on the unit space of G_n which normalizes the ample semi-group of G_n, that is, such that for any G-map s in the ample semi-group of G_n, $\phi^0 \circ s \circ \phi^{0-1}$ is also in the ample semi-group of G_n. Explicitly, for $n = 1$, the shift ϕ^0 on $\bar{\mathbb{Z}}$ sends u into u - 1 if u is finite and ∞ into ∞. For $n \geq 2$, ϕ^0 is given by $\phi^0 u = v$ where $v_i = u_{i-1}$. The shift ϕ^0 induces an automorphism ϕ of the locally compact groupoid G_n. Explicitly, ϕ is given by $\phi(u,v) = (\phi^0(u),\phi^0(v))$. We let $\mathbb{Z}$ act on G_n by $z \to \phi^z$ and form the semi-direct product $G_n \times_\phi \mathbb{Z}$ (see 1.1.7 and the beginning of the proof of 2.3.9). It is an r-discrete groupoid admitting the counting measures on the fibers as Haar system.

Finally, we define for each n the following subset of unit space of G_n. For $n = 1$, $O_1^0 = \bar{\mathbb{N}} = \mathbb{N} \cup \{\infty\}$. For $2 \leq n < \infty$, $O_n^0 = \{u \in \{0,1,\ldots,n-1\}^{\mathbb{Z}} : u_i = 0 \text{ for } i < 0\}$. It will be identified with $\{0,1,\ldots,n-1\}^{\mathbb{N}}$. For $n = \infty$, $O^0 = \{u \in G_\infty^0 : u_i = 0 \text{ for } i < 0\}$. It will be identified with $\{u \in \bar{\mathbb{N}}^{\mathbb{N}} : u_i = \infty \Longrightarrow u_j = \infty \text{ for every } j \geq i\}$. Each of these subsets O_n^0 is closed in G_n^0, hence is a compact space.

<u>2.1. Definition :</u> Let $n = 1,2,\ldots,\infty$. The <u>Cuntz groupoid O_n</u> is the reduction of the semi-direct product $G_n \times_\phi \mathbb{Z}$ to the closed subset O_n^0 of its unit space (identified with the unit space of G_n).

Let us spell out the algebraic structure of the groupoids O_n. For $n = 1$, $O_1 = \{(u,z) \in \bar{\mathbb{N}} \times \mathbb{Z} : u + z \in \bar{\mathbb{N}}\}$, where $\infty + z = \infty$. The multiplication is given by $(u,z)(u + z, z') = (u,z + z')$ and the inverse of (u,z) is $(u + z, -z)$. The range unit of (u,z) is u and its domain unit is $u + z$. For n greater than 1 but finite,

$$O_n = \{(u,v,z) \in \{0,1,\ldots,n-1\}^{\mathbb{N}} \times \{0,1,\ldots,n-1\}^{\mathbb{N}} \times \mathbb{Z} : u_i = v_{i-z} \text{ for all but finitely many } i\text{'s}\}.$$

The multiplication is given by $(u,v,z)(v,w,z') = (u,w,z + z')$ and the inverse of (u,v,z) is $(v,u,-z)$. The range unit of (u,v,z) is u and its domain unit is v. In the case $n = \infty$,

$$O_\infty = \{(u,v,z) \in O^0 \times O^0 \times \mathbb{Z} : u_i = v_{i-z} \text{ for all but finitely many } i\text{'s and } k(u) = k(v) = z \text{ in the case when } k(u) \text{ or } k(v) \text{ is finite}\}.$$

The multiplication and the inverse are given as above.

The next task is to determine the ample semi-group of the groupoids O_n. Let us first define the Cuntz inverse semi-group $\mathcal{O}_n$, introduced implicitly in the first section of [15]. The semi-group $\mathcal{O}_1$ is the bicyclic semi-group [11].

<u>2.2. Definition :</u> Let $n = 1,2,\dots,\infty$. The <u>Cuntz inverse semi-group $\mathcal{O}_n$</u> is the semi-group consisting of an identity 1, a zero element 0 and all the words in the letters p_i, q_i with $i = 1,\dots,n$, subject to the relations $q_j p_i = 0$ if $i \neq j$ and $q_i p_i = 1$.

Let us recall the notations of [15]. Let $n = 1,2,\dots,\infty$. Given $k \in \mathbb{N}$, let $W_k^n = \{\alpha = (j_1,\dots,j_k) : j_i \in \{0,1,\dots,n-1\}\}$, $W_0^n = \{0\}$ and $W_\infty^n = \bigcup_{k=0}^{\infty} W_k^n$. For $\alpha = (j_1,\dots,j_k) \in W_\infty^n$, let $p_\alpha = p_{j_1} p_{j_2} \dots p_{j_k}$ and $q_\alpha = q_{j_k} q_{j_{k-1}} \dots q_{j_1}$. Then, it is shown in [15] (lemma 1.3) that any word in $p_i q_i$ may be uniquely written $p_\alpha q_\beta$ with $\alpha,\beta \in W_\infty^n$. The semi-group $\mathcal{O}_n$ is an inverse semi-group with $(p_\alpha q_\beta)^{-1} = p_\beta q_\alpha$, $1^{-1}=1$ and $0^{-1} = 0$. Its set of idempotent elements is $\mathcal{O}_n^0 = \{p_\alpha q_\alpha : \alpha \in W_\infty^n\}$ $\{0,1\}$. The order on $\mathcal{O}_n^0$ is $(i_1,\dots,i_k) \leq (j_1,\dots,j_\ell)$ iff $k \geq \ell$ and $i_m = j_m$ for $m = 1,\dots,\ell$.

The compact open G-sets of $G_n \times_\phi \mathbb{Z}$ are of the form $S \times \{z\}$ where S is a compact open G-set of G_n and $z \in \mathbb{Z}$. Therefore, the compact open G-sets of O_n are of the form $\{(u,z) : u \in S\}$ where $z \in \mathbb{Z}$ and S is a compact open subset of $\overline{\mathbb{N}}$ such that $S + z \subset \overline{\mathbb{N}}$ in the case $n = 1$ and of the form $\{(u,v,z) \in O_n^0 \times O_n^0 \times \mathbb{Z} : (u,v(-z)) \in S\}$ where S is a compact open G-set of G_n and $[v(-z)]_i = v_{i-z}$, in the case $n \geq 2$. In particular, let us define, for every $n = 1,2,\dots,\infty$ and every $\alpha,\beta \in W_\infty^n$ the following compact open G-sets of O_n. For $n = 1$, $S(\alpha,\beta) = \{(u,\ell(\beta) - \ell(\alpha)) : u \in [\ell(\alpha),\infty]\}$, where the length $\ell(\alpha)$ of α is k if α is in W_k^n. For $n \geq 2$, $S(\alpha,\beta) = \{((\alpha,u),(\beta,u),\ell(\alpha) - \ell(\beta)) : u \in O_n^0\}$.

<u>2.3. Proposition :</u> Let $n = 1,2,\dots,\infty$. The map which sends $p_\alpha q_\beta$ into $S(\alpha,\beta)$, 0 into $\emptyset$ and 1 into O_n^0 is an isomorphism of the inverse semi-group $\mathcal{O}_n$ into the ample semi-group of the groupoid O_n. Its image, which will also be denoted $\mathcal{O}_n$, generates the ample semi-group in the sense that

(i) every compact open set of O_n^0 may be written as the difference $A \setminus B$ of two sets A and B which are both a finite disjoint union of elements of $\mathcal{O}_n^0$.

(ii) every compact open G-set of O_n may be written as a finite union $\bigcup_1^\ell E_i S_i F_i$ where (E_i) and (F_i) are two families of disjoint compact open sets in O_n^0 and the S_i's are in $\mathcal{O}_n$.

Proof : The map is clearly one-to-one. In order to show that it is a homomorphism, it suffices to consider the generators p_i and q_i's. Let us define, for α and β in W^n_∞, $P_\alpha = S(\alpha,0)$ and $Q_\beta = S(0,\beta)$. Thus, $S(\alpha,\beta) = P_\alpha Q_\beta$. We write P_i instead of $P(i)$. Then, the following relations are satisfied : $Q_j P_i = \emptyset$ if $i \neq j$, $Q_i P_i = O^0_n$, $P_i P_j = P_{(i,j)}$ and $Q_i Q_j = Q_{(j,i)}$. Therefore, the map is an isomorphism of $\mathcal{O}_n$ into the ample semi-group of O_n.

The image of the idempotent element $p_\alpha q_\alpha$ is the interval $[\ell(\alpha),\infty]$ in the case $n = 1$ and the cylinder set $Z(\alpha)$ of O^0_n in the case $n \geq 2$. In the case $n - 1$, the assertion (i) is clear. It is also clear in the case $2 \leq n < \infty$ since every compact open set of $\{0,1,\dots,n-1\}^{\mathbb{N}}$ is a finite disjoint union of cylinder sets $Z(\alpha)$. In the case $n = \infty$, it suffices to check that the sets $A \setminus B$, where both A and B are unions of cylinder sets $Z(\alpha)$ form a base for the topology of O^0_∞. This is immediate from the definition of the topology of O^0_∞.

The last assertion is also clear. For example, in the case $n \geq 2$, the G-set $\{(u,v,0) : (u,v) \in S\}$ where S corresponds to the transposition $(\alpha,u) \to (\beta,u)$, with $\alpha,\beta \in W^n_k$, belongs to $\mathcal{O}_n$.

Q.E.D.

2.4. Remark : The groupoid O_n has the property of having its ample semi-group generated by the inverse semi-group $\mathcal{O}_n$. Two questions arise for which we have no answer. Given an inverse semi-group $\mathcal{G}$, does there exist an r-discrete groupoid G whose ample semi-group of compact open G-sets is generated by $\mathcal{G}$ and covers G ? What kind of uniqueness can we expect ?

Let us note that the realization of an inverse semi-group $\mathcal{G}$ as a semi-group of G-sets introduces an extra structure on $\mathcal{G}$ and embeds its idempotent elements into a Boolean algebra and allows the definition of a partial addition : S and T can be added provided that $r(S) \cap r(T) = \emptyset$ and $d(S) \cap d(T) = \emptyset$, then S + T is the union of S and T. For example, we have introduced in the case $2 \leq n < \infty$ the relation $\sum_{i=1}^{n} P_i Q_i = I$.

2.5. Proposition :

(i) For every $n = 1,2,\dots,\infty$, the groupoid O_n is amenable.

(ii) For $n = 1$, $\mathbb{N}$ is an open invariant set for O_1. The reduction of O_1 to $\mathbb{N}$ is the transitive groupoid on $\mathbb{N}$ and the reduction of O_1 to $\{\infty\}$ is the group $\mathbb{Z}$.

(iii) For $n \geq 2$, the groupoid O_n is minimal.

(iv) For every $n = 1,2,\ldots,\infty$, the groupoid O_n has a base of compact open G-sets and is of infinite type.

Proof :

(i) We have constructed O_n as the reduction of a semi-direct product. We may apply 2.3.7 and 2.3.9.

(ii) The open subset of $\mathbb{N}$ of $\overline{\mathbb{N}}$ is clearly invariant. We may define the isomorphism of $O_{1|\mathbb{N}}$ onto $\mathbb{N}\times\mathbb{N}$ which sends (u,z) into $(u,u+z)$. The isotropy group of O_1 at $\{\infty\}$ is $\mathbb{Z}$.

(iii) The groupoid O_n induces the equivalence relation $\sim$ on its unit space, where $u \sim v$ iff there exists $z \in \mathbb{Z}$ such that $u_i = v_{i-z}$ for all but finitely many i's. Hence every orbit meets every cylinder set $Z(\alpha)$, where $\alpha \in W_n^\infty$ for $2 \leq n < \infty$, and every cylinder set $Z(\alpha)\setminus\bigcup_1^k Z(\beta_j)$, where $\alpha,\beta_j \in W_\infty^\infty$ for $n = \infty$. This shows that every orbit is dense.

(iv) The G-sets SE, where $S \in \mathcal{O}_n$ and E is a compact open set in O_n^0 constitute a base for the topology of O_n. Since, in $\mathcal{O}_n$, p_iq_i is equivalent to 1, the groupoid O_n is of infinite type.

Q.E.D.

Let us recall the definition of a representation of an inverse semi-group on a Hilbert space given by B. Barnes in [1], page 363.

.6. Definition : A representation of an inverse semi-group $\mathcal{G}$ on a Hilbert space H is an inverse semi-group homomorphism of $\mathcal{G}$ into an inverse semi-group of partial isometries of H.

Let V be a representation of the inverse semi-group $\mathcal{O}_n$, $n = 1,2,\ldots,\infty$. The images $S_i = V(p_i)$ of the generators p_i are isometries with mutually orthogonal ranges. Conversely, any sequence (S_i) $i = 1,\ldots,n$ of isometries with mutually orthogonal ranges defines a unique representation V of $\mathcal{O}_n$ such that $V(p_i) = S_i$ for every $i = 1,\ldots,n$.

In the case $2 \leq n < \infty$, we require that $\sum_{i=1}^{n} S_i S_i^* = 1$.

<u>2.7. Proposition :</u> Let $n = 1,2,\ldots,\infty$. There is a bijective correspondence between the representations V of $\mathcal{O}_n$ on separable Hilbert spaces, such that $\sum_{1}^{n} V(p_i)V(q_i) = 1$ in the case $2 \leq n < \infty$, and the representations of $C^*(O_n)$ on separable Hilbert spaces.

<u>Proof :</u> Suppose that L is a representation of $C^*(O_n)$ Then, by 2.1. 20, it gives by restriction a representation of the ample semi-group of O_n, hence a representation of $\mathcal{O}_n$. In the case $2 \leq n < \infty$, the relation $\sum_{i=1}^{n} E_i = 1$, where E_i is the characteristic function of the cylinder set Z(i), holds in $C^*(O_n)$ and gives the relation $\sum_{i=1}^{n} S_i S_i^* = 1$.

Conversely, suppose that V is a representation of $\mathcal{O}_n$ such that, in the case $2 \leq n < \infty$, $\sum_{i=1}^{n} S_i S_i^* = 1$, where $S_i = V(P_i)$. Its restriction to the set $\mathcal{O}_n^0$ of idempotent elements, which will be denoted M, is a monotone projection-valued function, taking the value 0 at 0 and the value 1 at **1** . It is finitely additive in the case $2 \leq n < \infty$, because of the relation $\sum_{i=1}^{n} S_i S_i^* = 1$. We will extend it to a finitely additive projection-valued measure on the Boolean algebra $\mathcal{B}_n$ of compact open subsets of O_n^0. In the case n = 1, any compact open subset of $\overline{\mathbb{N}}$ is a finite disjoint union of difference of elements of $\mathcal{O}_1^0$. Thus, if $A = \bigcup_{i=1}^{\ell} B_i \setminus C_i$ with T_i, $C_i \in \mathcal{O}_1^0$ and $C_i \subset B_i$, we define $M(A) = \sum_{i=1}^{\ell} M(B_i) - M(C_i)$. Because the order of $\mathbb{N}$ is total, M(A) is well defined and M is finitely additive. In the case $2 \leq n < \infty$,any compact open subset of $\{0,1,\ldots,n-1\}^{\mathbb{N}}$ is a finite disjoint union of elements of $\mathcal{O}_n^0$. Thus, if $A = \bigcup_{i=1}^{\ell} B_i$, with $B_i \in \mathcal{O}_n^0$, we define $M(A) = \sum_{i=1}^{\ell} M(B_i)$. This is well defined and additive because M is additive on $\mathcal{O}_n^0$. In the case $n = \infty$, we first extend M to the elements of $\mathcal{O}_\infty$ which are a finite disjoint union of elements of $\mathcal{O}_n^0$. Since every element A of $\mathcal{O}_\infty$ is the difference of two such elements, say $A = B \setminus C$ with $C \subset B$, we may define $M(A) = M(B)-M(C)$. One shows as in the case n = 1 that M is well defined and additive.

Having extended M to the Boolean algebra $\mathcal{B}_n$, we may extend V to a representation of the ample semi-group of O_n. We know from 2.3.(iii) that every compact open G-set of O_n may be written as a finite union $S = \bigcup_{1}^{\ell} E_i S_i F_i$ where (E_i) and (F_i) are two families of disjoint elements of $\mathcal{B}_n$ and the S_i's are in $\mathcal{O}_n$. We define

$V(S) = \sum_{1}^{\ell} M(E_i)V(S_i)M(F_i)$. It is a partial isometry and it does not depend on the way S has been written. Moreover, it is an inverse semi-group homomorphism.

The pair (V,M) is a covariant representation of O_n (cf.2.1.20) and can be extended to a representation of $C^*(O_n)$. Explicitly, every $f \in C_c(O_n)$ may be written $f = \sum_{1}^{\ell} h_i \chi_{S_i}$ where $h_i \in C_c(O_n^0)$ and S_i is a compact open G-set of O_n. We define $L(f) = \sum_{1}^{\ell} M(h_i)V(S_i)$. A computation similar to one given in the proof of 2.4.15 shows that L(f) is well defined. Moreover, the map L so defined is a representation of $C_c(O_n)$ continuous for the inductive limit topology. Since r-discrete groupoids with Haar system have sufficiently many non-singular Borel G-sets, we know from 2.1.22 that L extends to a representation of $C^*(O_n)$.

Q.E.D.

2.8. Remarks :

(i) In order to study the representations of an inverse semi-group $\mathcal{G}$ on a Hilbert space, B. Barnes makes use in [1] and [2] of its Banach $*$-algebra $\ell^1(\mathcal{G})$. He shows in particular that $\ell^1(\mathcal{G})$ has a faithful representation. The example of $\mathcal{O}_n$ suggests another approach. One can try first to realize the inverse semi-group as a generating subsemi-group of the ample semi-group of a groupoid G and then define the C^*-algebra of $\mathcal{G}$ as $C^*(G)$. The example of the bicyclic semi-group O_1 is studied in [1] (section 7). The description of its irreducible representations given there can also be obtained from 2.5. (ii).

(ii) The C^*-algebra $C^*(O_n)$ is generated by the isometries P_i, $i = 1,\ldots,n$. Indeed these isometries generate $\mathcal{O}_n$ as an inverse semi-group. Moreover $\mathcal{O}_n^0$ generates the Boolean algebra $\mathcal{B}_n$ of compact open subsets of O_n^0. Therefore, the C^*-algebra generated by the P_i's contains $C_c(O_n)$. It must be $C^*(O_n)$. Thus $C^*(O_n)$ is, for $n \geq 2$, one of the C^*-algebras studied by J. Cuntz in [15]. It is shown there that such an algebra is simple. We can prove it directly. Indeed the groupoid O_n is amenable, minimal and essentially principal (definition 2.4.3). Hence we may apply proposition 2.4.6.

We have seen (1.1.7) that the semi-direct product $G_n \times_\phi \mathbb{Z}$ has a natural cocycle

$c_n \in Z^1(G_n \times_\phi \mathbb{Z}, \mathbb{Z})$, namely the cocycle given by $c_n(x,z) = z$. Its restriction to the reduction $G_n \times_\phi \mathbb{Z}\,|0_n^0$ is still a cocycle. Explicitly, for n=1, $c_1 \in Z^1(0_1, \mathbb{Z})$ is defined by $c_1(u,z) = z$ and for $n \geq 2$, $c_n \in Z^1(0_n, \mathbb{Z})$ is defined by $c_n(u,v,z) = z$. We may observe that the "fixed point" groupoid $c_n^{-1}(0)$ bears some resemblance with G_n. Indeed, for n = 1, $c_n^{-1}(0)$ is the unit space $\overline{\mathbb{N}}$ of 0_1. For $2 \leq n < \infty$, $c_n^{-1}(0)$ is the Glimm groupoid given by the equivalence relation $u \sim v$ iff $u_i = v_i$ for all but finitely many i's on $\{0,1,\ldots,n-1\}^{\mathbb{N}}$. For $n = \infty$, $c_\infty^{-1}(0)$ is the AF groupoid given by the equivalence relation $u \sim v$ iff $k(u) = k(v)$ and $u_i = v_i$ for all but finitely many i's on 0_∞^0. Its dimension group is the lexicographical direct sum $\sum_{i \in \mathbb{N}} \mathbb{Z}$ and its dimension range is the segment $[0,1]$, where $1 = (1,0,0,\ldots)$.

The following result, due to Olesen and Pedersen ([58]), is interesting because it exhibits the different behavior of the 0_n groupoids, in comparison to the AF groupoids, with respect to KMS measures. The definition of (c,β) KMS measures has been given in 1.3.15.

We will make use of the relation $\frac{d\mu \cdot s^{-1}}{d\mu}(u) = D^{-1}(us)$, where D is the modular function of μ and s is a G-map, established in 1.3.18. (iii) and 1.3.20.

__2.9. Proposition :__ Let $n = 1,2,\ldots,\infty$ and let $c_n \in Z^1(0_n, \mathbb{Z})$ be as above. Then

(i) if n = 1, there are no (c_1,β)-KMS probability measures for $\beta > 0$ and there exists a unique (c_1,β)-KMS probability measure for $\beta \leq 0$;

(ii) if $n \geq 2$, there are no (c_n,β)-KMS probability measures for $\beta \neq \log n$ and there exists a unique (c_n,β)-KMS probability measure for $\beta = \log n$.

__Proof :__

(i) Since $d^{-1}(u) = \{(v,u-v) : v \in \mathbb{N}\}$, if u is finite and $d^{-1}(\infty) = \{(\infty,z) : z \in \mathbb{Z}\}$, $\mathrm{Min}(c_1)$, which is the set of units u such that the restriction of c_1 to $d^{-1}(u)$ is non-negative, is empty, while the set $\mathrm{Max}(c_1)$ of units u such that the restriction of c_1 to $d^{-1}(u)$ is non positive is $\{0\}$. Therefore there is no ∞ KMS measure and the point mass at 0 is the unique $-\infty$ KMS probability measure.

Suppose that μ is a KMS probability measure on $\mathbb{N}$ at a finite β. Let Q be the G-set $\{(u,1) : u \in \mathbb{N}\}$ and let q be the corresponding G-map. For every compact open subset A of $\overline{\mathbb{N}}$, the following equalities hold :

$$\mu(A\cdot q) = \int \chi_A(u)\, d(\mu\cdot q^{-1})(u)$$
$$= \int \chi_A(u)\, D^{-1}(uq)\, d\mu(u)$$
$$= e^{\beta}\ \ \mu(A)\ .$$

In particular, for every $i \in \mathbb{N}$, $\mu\{i+1\} = e^{\beta}\ \mu\{i\}$. Since μ is required to be a probability measure, this is possible for $\beta \leq 0$ only. Then μ is uniquely defined by $\mu\{i\} = \dfrac{e^{\beta i}}{1 - e^{\beta}}$ if $\beta < 0$ and by $\mu\{\infty\} = 1$ for $\beta = 0$.

(ii) Suppose $2 \leq n < \infty$. Then $\mathrm{Min}(c_n) = \mathrm{Max}(c_n) = \emptyset$ because $d^{-1}(u) = \{(v,u,z) : v \sim u \text{ and } z \in \mathbb{Z}\}$ and there are no KMS measures at infinity. Let μ be a KMS probability measure on O_n^0 at a finite β. We know (1.3.16) that μ is $c_n^{-1}(0)$ invariant ; however the Glimm groupoid $c_n^{-1}(0)$ has a unique invariant probability measure, because of the structure of its dimension range. This is the measure μ_n defined by $\mu_n(Z(\alpha)) = n^{-\ell(\alpha)}$ for every $\alpha \in W_n^{\infty}$. Since $\mu_n(A\cdot p_\alpha q_\beta) = n^{(\ell(\alpha)\ -\ \ell(\beta))}\ \mu_n(A)$ for every compact open set A and every pair (α,β) in W_n^{∞}, the modular function of μ_n with respect to O_n^{∞} is $D_n(u,v,z) = n^{-z} = \exp(-\log n\ c(u,v,z))$. Thus β must be equal to $\log n$.

(iii) Suppose $n = \infty$. Since $d^{-1}(u)\ \{(v,u,k(v) - k(u)) : v \sim u\}$ if $k(u)$ is finite and $d^{-1}(u) = \{(v,u,z) : v \sim u,\ z \in \mathbb{Z}\}$ if $k(u)$ is infinite, $\mathrm{Max}(c_\infty)$ is always empty and $\mathrm{Min}(c_\infty) = \{\infty\}$ where ∞ denotes the sequence $(\infty,\infty,\ \ldots)$. Thus there is no KMS measures at $\beta = -\infty$ and the point mass at $\{\infty\}$, μ_∞, is the only KMS probability measure at $\beta = \infty$. There cannot be any KMS probability measures at a finite β because μ_∞, which is the only probability measure invariant under $c_\infty^{-1}(0)$, is not quasi-invariant under O_∞.

Q.E.D.

.10. Remark : If G is the groupoid of a transformation group (U,S), where S is a subgroup of $\mathbb{R}$, the only possible KMS probability measures for the cocycle $c(u,s) = s$ are invariant probability measures.

Appendix.

THE DIMENSION GROUP OF THE GICAR ALGEBRA

We have seen that AF groupoids are classified by their dimension ranges (3.1.13. ii). Therefore, the computation of the dimension range is the essential step in the study of AF groupoids. The problem can be split into two parts, first the computation of the dimension group, second the determination of the dimension range as an upward directed hereditary subset of the positive cone of the dimension group. The first part is more difficult. An Elliott group, that is, the dimension group of an AF groupoid, is usually given as an inductive limit. Although some information can be read off from the corresponding diagram, in particular its ideal structure (see [8]), and one can decide when two diagrams give isomorphic dimension groups, it is of great interest to have an intrinsic definition of the dimension group. This is why this computation is included here, which is probably known to others.

Let us recall the definitions : the CAR groupoid is $CAR = \{(u,v) \in U \times U : u_i = v_i \text{ a.e.}\}$ where $U = \{0,1\}^{\infty}$ the GICAR groupoid is the subgroupoid $GICAR = c^{-1}(0)$ where $c(u,v) = \sum u_i - v_i$. Its ample semi-group consists of the G-maps of CAR which "preserve the number of particles". This means $\sum_1^j (u.s)_i = \sum_1^j u_i$ for j large enough. Writing GICAR as an inductive limit, we obtain the following inductive system for its dimension group :

$$\cdots \longrightarrow \mathbb{Z}^n \xrightarrow{\begin{bmatrix} 1 & 0 \\ 1 & 1 \\ 0 & 1 \end{bmatrix}} \mathbb{Z}^{n+1} \longrightarrow \cdots \quad n = 1,2, \ldots$$

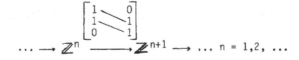

<u>Proposition</u> : The dimension group of GICAR is $\mathbb{Z}[t]$ = {polynomials in t with integer coefficients} with usual addition and order $f > 0$ iff $f(t) > 0$ for any $t \in]0,1[$.

<u>Proof</u> :

(a) We first observe that for a given n and $p = 0,\dots,n$,

$$(1)\quad t^p = \sum_{k=0}^{n} \binom{n-p}{k-p} t^k (1-t)^{n-k} \quad \text{and}$$

$$(1-t)^p = \sum_{k=0}^{n} \binom{n-p}{k} t^k (1-t)^{n-k}.$$

(b) We introduce $e_k^n = t^k (1-t)^{n-k}$ for $n = 1,2,\dots$ and $k = 0,\dots,n$ and remark that

(2) the e_k^n's generate $\mathbb{Z}[t]$,

(3) for fixed n, the e_k^n's are linearly independent,

(4) $e_k^n = e_k^{n+1} + e_{k+1}^{n+1}$ $n = 1,2,\dots$ and $k = 0,1,\dots,n$.

Hence, as a group $\mathbb{Z}[t]$ is the limit of the inductive system. Let us determine the order the system induces on $\mathbb{Z}[t]$. By definition, $f > 0$ iff for sufficiently large n, the coefficients λ_k of the expansion $f = \sum_{k=0}^{n} \lambda_k t^k (1-t)^{n-k}$ are non-negative.

(c) We show that $f > 0$ and $g > 0$ imply $fg > 0$. Indeed,

$$f = \sum_{k=0}^{m} \lambda_k t^k (1-t)^{m-k} \quad \lambda_k \geq 0,\ k = 0,\dots,m,$$

$$g = \sum_{\ell=0}^{n} \mu_\ell t^\ell (1-t)^{n-\ell} \quad \mu_\ell \geq 0,\ \ell = 0,\dots,n, \text{and}$$

$$fg = \sum_{j=0}^{m+n} \Big(\sum_{k+\ell=j} \lambda_k \mu_\ell\Big) t^j (1-t)^{m+n-j} \quad \text{with}$$

$$\sum_{k+\ell=j} \lambda_k \mu_\ell \geq 0, \dots, m+n.$$

(d) Let $f \in \mathbb{Z}[t]$ such that $f(t) > 0$ for $t \in [0,1]$, then $f > 0$. We write

$$f = \sum_{p=0}^{m} a_p t^p \text{ and } f_n = \sum_{p=0}^{m} a_p \, t \, \frac{t - \frac{1}{n}}{1 - \frac{1}{n}} \cdots \frac{t - \frac{p-1}{n}}{1 - \frac{p-1}{n}}, \; n \geq m.$$

Since f_n converges to f uniformly on $[0,1]$, for n sufficiently large, $f_n(t) > 0$ for $t \in [0,1]$. Then,

$$f = \sum_{p=0}^{m} a_p t^p = \sum_{p=0}^{m} a_p \sum_{k=0}^{n} \binom{n-p}{k-p} t^k (1-t)^{n-k}$$

$$= \sum_{k=0}^{n} \left[\sum_{p=0}^{m} a_p \binom{n-p}{k-p} \right] t^k (1-t)^{n-k}, \text{ and}$$

$$\sum_{p=0}^{m} a_p \binom{n-p}{k-p} = \binom{n}{k} \sum_{p=0}^{m} a_p \frac{k(k-1)\ldots(k-p+1)}{n(n-1)\ldots(n-p+1)} = \binom{n}{k} f_n\left(\frac{k}{n}\right) > 0.$$

(e) We conclude that for a non zero $f \in \mathbb{Z}[t]$, $f > 0$ iff $f(t) > 0$ for $t \in]0,1[$. The condition is clearly necessary. To show that it is sufficient, write $f = t^m g(1-t)^n$ with $g(t) > 0$ for $t \in [0,1]$. By (d), $g > 0$ and by (c), $f > 0$.

REFERENCES

1. B. Barnes, Representations of the ℓ^1-algebra of an inverse semi-group, Trans. Amer. Math. Soc. 218 (1976), 361-395.
2. B. Barnes, Representations of the ℓ^1-algebra of an inverse semi-group having the separation property, Glasgow Math. J. 18 (1977), 131-143.
3. R. Blattner, Positive Definite Measures, Proc. Amer. Math. Soc. 14 (1963), 423-428.
4. N. Bourbaki, Topologie générale, Chap. 9, Hermann, Paris, 1958.
5. N. Bourbaki, Intégration, Chap. 1-6, Hermann, Paris, 1965.
6. N. Bourbaki, Intégration, Chap. 7-8, Hermann, Paris, 1963.
7. N. Bourbaki, Intégration, Chap. 9, Hermann, Paris, 1963.
8. O. Bratteli, Inductive limits of finite-dimensional C^*-algebras, Trans. Amer. Math. Soc. 171 (1972) 195-234.
9. O. Bratteli, Crossed products of UHF algebras by product type actions, Duke Math J. 46 (1979).
0. R. Brown and J. Hardy, Topological groupoids : I Universal constructions, Math. Nach 71 (1976).
1. A. H. Clifford and G.B. Preston, The algebraic theory of semi-groups, Vol. 1, Math. Surveys, no. 7, Amer. Math. Soc., Providence, R.I., 1961.
2. F. Combes, Poids sur une C^*-algèbre, J. Math. pures et appl., 47 (1968) 57-100.
3. A. Connes, Une classification des facteurs de type III, Ann. Scient. Ec. Norm. Sup., 6 (1973), 133-252.
. A. Connes, Sur la théorie non commutative de l'intégration, Springer Lecture Notes in Math. 725 (1979), 19-143.
. J. Cuntz, Simple C^*-algebras generated by isometries, Commun. Math. Phys. 57 (1977), 173-185.
. J. Dixmier, Algèbres quasi-unitaires, Comment. Math. Helv. 26 (1952), 275-322.
. J. Dixmier, Sous-anneaux abéliens maximaux dans les facteurs de type fini, Ann. of Math. 59 (1954), 279-286.
. J. Dixmier, On some C^*-algebras considered by Glimm, J. Funct. Anal. 1 (1967), 182-203.

19. J. Dixmier, Les C^*-algèbres et leurs représentations, Gauthier-Villars, Paris, 1969.

20. S. Doplicher, D. Kastler and D. W. Robinson, Covariance algebras in field theory and statistical mechanics, Commun. Math. Phys. 3 (1966), 1-28.

21. H. A. Dye, On groups of measure preserving transformations I, Amer. J. Math. 81 (1959), 119-159.

22. H. A. Dye, On groups of measure preserving transformations II, Amer. J. Math. 85 (1963), 551-576.

23. E. Effros, Transformation groups and C^*-algebras, Ann. of Math. 81 (1965), 38-55.

24. E. Effros and F. Hahn, Locally compact transformation groups and C^*-algebras, Memoirs Amer. Math. Soc. 75 (1967).

25. E. Effros, D. Handelman and C.-L. Shen, Dimension groups and their affine representations, preprint.

26. C. Ehresmann, Catégories topologiques et catégories différentiables, Colloque de géométrie différentiable globale, Bruxelles, 1958.

27. G. Elliott, On the classification of inductive limits of sequences of semi-simple finite-dimensional algebras, J. of Algebra 38 (1976), 29-44.

28. G. Elliott, On totally ordered groups, preprint, Copenhagen, March 1978.

29. G. Emch, Algebraic methods in statistical mechanics and quantum field theory, Wiley Interscience, New York, 1972.

30. P. Eymard, Moyennes invariantes et représentations unitaires, Springer Lecture Notes in Math. 300, 1972.

31. J. Feldman and C. Moore, Ergodic equivalence relations, cohomology and von Neumann algebras, I and II, Trans. Amer. Math. Soc. 234, no 2 (1977), 289-359.

32. J. Fell, Weak containment and induced representations of groups. II, Trans. Amer. Math. Soc. 110 (1964), 424-447.

33. J. Fröhlich, The pure phases (harmonic functions) of generalized processes or : Mathematical Physics of phase transition and symmetry breaking, Bull. Amer. Math. Soc. 84 (1978) 165-193.

34. L. Gårding and A. Wightmann, Representations of the anticommutation relations, Proc. Nat. Acad. Sci. USA 40 (1954), 617-621.

35. J. Glimm, On a certain class of operator algebras, Trans. Amer. Math. Soc. 95 (196), 318-340.

36. J. Glimm, Locally compact transformation groups, Trans. Amer. Math. Soc. 141 (1961), 124-138.

37. J. Glimm, Families of induced representations, Pacific J. Math. 12 (1962), 885-911.

38. E. Gootman and J. Rosenberg, The structure of crossed product C^*-algebras : A proof of the generalized Effros-Hahn conjecture, Invent. Math. 52 (1979), 283-298.

39. P. Green, C^*-algebras of transformation groups with smooth orbit space, Pacific J. Math., 72 (1977), 71-97.

40. P. Green, The local structure of twisted covariance algebras, Acta Math. 140 (1978), 191-250.

41. F. Greenleaf, Invariant means on topological groups, Van Nostrand Reinhold, New York, 1969.

42. A. Guichardet, Une caractérisation des algèbres de von Neumann discrètes, Bull. Soc. Math. France 89 (1961), 77-101.

43. P. Hahn, Haar measure and convolution algebra on ergodic groupoids, Ph. D. Thesis, Harvard University, May 1975.

44. P. Hahn, Haar measure for measure groupoids, Trans. Amer. Math. Soc. 242, no. 519 (1978), 1-33.

45. P.Hahn, The regular representation of measure groupoids, Trans. Amer. Math. Soc. 242, no. 519 (1978), 35-72.

46. K. Hofmann and F. Thayer, Approximately finite dimensional C^*-algebras, preprint, Tulane University, 1976.

47. B. E. Johnson, An introduction to the theory of centralizers, Proc. London Math. Soc. 14 (1964) 299-340.

48. B.E. Johnson, Cohomology in Banach algebras, Memoirs Amer. Math. Soc. 127 (1972).

49. W. Krieger, On the Araki-Woods asymptotic ratio set and non singular transformations, Lecture Notes in Mathematics 160, Berlin 1970.

50. J. Cuntz and W. Krieger, A Class of C^*-algebras and Topological Markov chains, preprint (1979).

51. W. Krieger, On dimension functions and topological Markov chains, preprint (1979).

52. W. Krieger, On a dimension for a class of homomorphism groups, preprint.

53. G. W. Mackey, Ergodic theory and virtual groups, Math. Ann. 166 (1966), 187-207.

54. F. Murray and J. von Neumann, On rings of operators, Ann. of Math. (2) 37 (1936), 116-229.

55. C. Moore, Invariant measures on product spaces, Fifth Berkeley Symposium, II part 2 (1967), 447-459.

56. D. Olesen, Inner $*$ automorphisms of simple C^*-algebras, Commun. Math. Phys. 44 (1975), 175-190.

57. D. Olesen, G. K. Pedersen and E. Stormer, Compact abelian groups of automorphisms of simple C^*-algebras, Inventiones Math. 39 (1977), 55-64.

58. D. Olesen and G. K. Pedersen, Application of the Connes spectrum to C^*-dynamical systems, J. Funct. Anal. 30 (1978), 179-197.

59. D. Olesen and G. K. Pedersen, Some C^*-dynamical systems with a single KMS state, Math. Scand. 42 (1978), 111-118.

60. G. K. Pedersen, C^*-algebras and their automorphism groups, Academic Press, London, 1979.

61. A. Ramsay, Virtual groups and group actions, Advances in Math. 6 (1971), 253-322.

62. G. Rauzy, Répartitions Modulo 1, preprint, Marseille, 1977.

63. M. Rieffel, Induced representations of C^*-algebras, Advances in Math. 13 (1974) 176-257.

64. S. Sakai, C*-algebras and W*-algebras, Springer, Berlin, 1971.

65. S. Sakai, Recent developments in the theory of unbounded derivations in C^*-algebras, Proceedings of the US-Japan seminar on C^*-algebras and their application to theoretical physics, U.C.L.A., April 1977, Springer Lecture Notes in Math. 650, 85-122.

66. J.-L. Sauvageot, Idéaux primitifs de certains produits croisés, Math. Ann. 231 (1977), 61-76.

67 A. K. Seda, A continuity property of Haar systems of measures, Ann. Soc. Sci. Bruxelles 89 IV (1975), 429-433.

68. A. K. Seda, Haar systems for groupoids, Proc. Royal Irish Acad. 76, Sec. A, no. 5, (1976), 25-36.

69. S. Strătilă and D. Voiculescu, Representations of AF-algebras and of the group $U(\infty)$, Springer Lecture Notes in Math. 486, 1975

70. H. Takai, On a duality for crossed products of C^*-algebras, J. Funct. Anal. 19 (1975), 24-39.

71. H. Takai, The quasi-orbit space of continuous C^*-dynamical systems, Trans. Amer. Math. Soc. 216 (1976), 105-113.

72. M. Takesaki, Covariant representations of C^*-algebras and their locally compact automorphism groups, Acta Math. 119 (1967), 273-303.

73. M. Takesaki, Tomita's theory of modular Hilbert algebras and its applications, Lecture Notes in Math. 128, Springer, Berlin, 1970.

74. V. S. Varadarajan, Geometry of quantum theory, Vol. 2, Van Nostrand-Reinhold, New York, 1970.

75. J. Westman, Harmonic analysis on groupoids, Pacific J. Math. 27 (1968), 621-632.

76. J. Westman, Cohomology for ergodic groupoids, Trans. Amer. Math. Soc. 146 (1969), 465-471.

77. J. Westman, Non transitive groupoid algebras, University of California at Irvine, 1967.

78. J. Westman, Ergodic groupoid algebras and their representations, University of California at Irvine, 1968.

79. J. Westman, Groupoid theory in algebra, topology and analysis, University of California at Irvine, 1971.

80. G. Zeller-Meier, Produits croisés d'une C^*-algèbre par un groupe d'automorphismes, J. Math. pures et appl. 47 (1968), 101-239.

81. R. Zimmer, Amenable ergodic group actions and an application to Poisson boundaries of random walks, J. Funct. Anal. 27 (1978), 350-372.

82. R. Zimmer, On the von Neumann algebra of an ergodic action, Proc. Amer. Math. Soc. 66 (1977), 289-293.

83. R. Zimmer, Hyperfinite factors and amenable ergodic actions, Invent. Math. 41 (1977), 23-31.

NOTATION INDEX

(References are to the pages on which the symbols are defined).

GROUPOID ALGEBRAS (continued)

SUBJECT INDEX

(The first reference is to the page of these notes where the expression is defined ; the following references are to articles where a similar notion appears ; they are intended only to serve as a guide to the subject ; standard references to C^*-algebra theory are [19],[64] and [60]).